V-Bombs and Weathermaps

V-Bombs and Weathermaps

Reminiscences of
World War II

BROCK McELHERAN

McGill-Queen's University Press
Montreal & Kingston • London • Buffalo

© McGill-Queen's University Press 1995
ISBN 0-7735-1330-2

Legal deposit fourth quarter 1995
Bibliothèque nationale du Québec

Printed in Canada on acid-free paper

McGill-Queen's University Press is grateful
to the Canada Council for support of
its publishing program.

Canadian Cataloguing in Publication Data

McElheran, Brock, 1918–
V-bombs and weathermaps :
reminiscences of World War II
Includes bibliographical references and index.
ISBN 0-7735-1330-2
1. McElheran, Brock, 1918–   . 2. World War,
1939–1945 – Aerial operations, German.
3. Great Britain – History – Bombardment,
1944–1945. 4. World War, 1939–1945 – Personal
narratives, Canadian. I. Title.
D811.5.M34 1995   940.54'4943'092   C95-900238-3

Typeset in Palatino 10.5/13
by Caractéra production graphique, Quebec City

*Dedicated to the children, women, and men
whose courage and resolution triumphed over
Hitler's Vengeance weapons.*

HAVE WE FORGOTTEN?

Contents

Preface

Since the events described in this book took place, two generations have been born and educated – generations that have spent more time taming computers and breaking records than learning about the recent past. It is hoped that reading these pages will give some idea of what life was like in England during the attacks by Hitler's Vengeance weapons, the V-1 flying bomb or pilotless aircraft, and the V-2 high-altitude 4,000-MPH rocket.

I was privileged to live among the brave people of southern England from the mysterious arrival of the first robot aircraft until the end of the war and formed the highest opinion of their courage and dedication as well as their humour. Personal reminiscences are here combined with official facts and figures in an attempt to recreate the atmosphere of those tense but also exhilarating times.

The lighter passages must not be allowed to obscure the underlying seriousness of the book, indicated by the motto "Have We Forgotten?" This is a variation on the theme heard so often after the First World War, "Lest We Forget." I fear that the postwar generations have little idea of the suffering and gallantry of the civilian population in England during both wars, when the future of freedom hung in the balance.

In addition to accounts of life in London and its outlying areas during the V-bomb attacks, something is told here of the magnificent plastic-surgery hospital at East Grinstead, where the morale of the patients was given as much care as their faces. A section about the Battle of the Atlantic has been included. Most of these stories have not been told in print, as far as I know, and I hope that they may be of some interest, as well as disclosing what goes on in a weather forecaster's mind when the weather misbehaves.

From time to time statements about the general progress of the war are included, to serve as lighthouses for the reader whose memory is a trifle foggy. Chapter 2, in particular, reviews earlier attacks on Britain. It is surprising to learn the extent of the air raids in WWI, and the extent to which they have been forgotten.

References are made to the role of women during the war. Their heroic actions had a profound effect on all who witnessed them and doubtless contributed to the subsequent acceptance in Britain of a woman prime minister, the first to hold such a position in Europe.

The final chapter contrasts the justifiable recognition given to the men who served in both world wars with the skimpy acclaim accorded civilians – those employed in war work as well as those who, in the words of a London policeman, "just carried on."

The closing pages attempt to answer the question "Why not forget it all?" The book concludes with a challenge and a warning to everyone who hopes to survive in the twenty-first century.

For those who have read little about this or any other war, every effort has been made to avoid technical jargon. One exception remains: a ship hove to. None of the young people who assisted me had the slightest notion what that meant. But I couldn't risk offending any seafarers by saying "came to a stop." (Any land-lubber who has read this far is now well equipped to tackle chapter 22.)

A glossary has been added to help unscramble such initials as RCNVR and VAD.

The book was begun in 1944, soon after most of the incidents described herein. Extracts from that account are printed in special type. They are shortened but not rewritten, in case some long-suffering Ph.D. candidate has to write a thesis entitled "The Literary Style of the Late War Years." It even retains a few words now in disrepute, such as "slum" and "charwoman." However, the war ended before the book was finished, and I took a forty-five-year break to earn a living.

A word about the term "British." Canadians, like all other citizens of the Commonwealth, were legally British. They wore the same uniform as the home forces, except their buttons said "Canada" in small print, as did shoulder flashes in some branches. Drill, regulations, decorations, and ranks were the same. We were designed to be interchangeable with the UK personnel. Canadians fought under the Union Jack, in ships that wore the White Ensign, and flew aircraft marked by the British roundel. We were part of one huge force. However, when writing of the UK services, one usually calls them "British" as opposed to Canadian or Australian, etc. This problem of terminology has plagued writers of both wars.

In conclusion, may I thank those several people who acted as my eyes, my fingers, and my advisers – Joan Broadley, Christopher Emerson, Jennifer Jones, Juli Pomainville, Sally Skyrm, Elizabeth Avery, and my principal assistant, Kathleen Pauquette. The ladies of the Clerical Center of the State University of New York College at Potsdam, New York, headed by Mary Lauzon, performed prodigious feats of deciphering a typescript that resembled a wiring diagram soaked in mud.

But most of all, my deepest thanks to my beloved wife, Janie. She shared fogs and cramped quarters in Halifax and suffered the anguish of the wife of a serviceman under attack, she tramped around pavements in London and snow-covered battlefields in Flanders, she made sense out of reference books that were often vague and contradictory, and she helped fight every sentence until most were reasonably under control.

Brock McElheran
Potsdam, New York

Acknowledgments

The author wishes to express his deep gratitude to many people who provided specific information, especially the following:

Mr D. Ashby, Naval Historical Branch, The Admiralty, London, for information on Convoy HX289M.

Ms Caroline A. Baish, *Reader's Digest*, Pleasantville, NY, for locating two articles when given only the scantiest of clues.

Mr John P. Bennett, FRCS, Queen Victoria Hospital, East Grinstead, for material on the history of the hospital.

Mr D.G. Gage, Friends of the Canadian War Museum, for information concerning the German mine field off Halifax.

Mr Martin Hayes, Principal Librarian, Local Studies, South Eastern Divisional Library, Worthing, for details on the Cuckfield bomb.

Mr Hugh Halliday, Canadian War Museum, for providing naval photographs.

Mr R.J. Huse and Miss S. Breeden, North Eastern Divisional Library, Crawley, for details concerning the air raid of 9 July 1943 on East Grinstead.

Mr R.W. Mason, National Meteorological Library, Bracknell, for providing statistics of rainfall during the summer of 1944.

Mr J.E. Meacham, Librarian, *The Telegraph*, London, for information on the first flying bomb attacks.

Mr Michael Petty, Principal Librarian, Local Studies, Cambridgeshire Library, Cambridge, for information concerning air attacks in Cambridgeshire.

Ms Elizabeth Purvis, Senior Reference Librarian, Sevenoaks Library, for information about the night of 12–13 June, and accounts of life in "Bomb Alley."

Mr Joseph V. Waters, Founder Member, The Ragged School Museum Trust, Bow, London, for information about East London air raids in two wars.

Mr Alan Williams, Imperial War Museum, London, for researching and providing photographs.

Dr H. Bruce Williams, Director, Division of Plastic and Reconstructive Surgery, Montreal General Hospital, for checking medical details.

Abbreviations and Glossary

ABBREVIATIONS

BBC British Broadcasting Corporation

GI Government Issue. Nickname for an American soldier.

Met. o. Meteorological officer. Called aerological officer in the US navy.

RAF Royal Air Force of Great Britain.

RCAF Royal Canadian Air Force.

RCN Royal Canadian Navy. Officers wore straight stripes on sleeves as rank insignia.

RCNVR Royal Canadian Naval Volunteer Reserve. Civilians in peacetime, full-time sailors in war. Officers wore wavy stripes on sleeves as rank insignia; hence the "Wavy Navy."

RN Royal Navy. The British permanent force. Officers wore straight stripes on sleeves as rank insignia. Also means Registered Nurse.

RNVR Royal Naval Volunteer Reserve. Civilians in peacetime, full-time sailors in war. Officers wore wavy stripes on sleeves as rank insignia.

V-1 German flying bomb or pilotless aircraft. Also called doodlebug or buzz bomb.

From German *Vergeltungswaffe*, or "Vengeance weapon."

v-2 The second of Hitler's Vengeance weapons. High-altitude, high-speed rocket.

VAD Voluntary Aid Detachment. Volunteer nurses in the armed forces of Great Britain.

WAAC Women's Army Auxiliary Corps.

WAAF Women's Auxiliary Air Force.

WAVES Women accepted for volunteer emergency service, American women's navy.

WRNS Women's Royal Naval Service. The British women's navy.

WWI The Great War or First World War, 1914–18.

WWII The Second World War, 1939–45.

GLOSSARY

Antisubmarine vessel See *Escort vessel.*

Asdic Underwater antisubmarine device. The US navy calls it sonar.

Battle of the Atlantic The many naval actions in the North Atlantic Ocean, particularly against German submarines.

Battle of the V-bombs The attack on England by the v-1 flying bombs and the v-2 rockets, from 13 June 1944 to 29 March 1945.

Blitz The attacks by conventional bombers against Britain, from September 1940, to May 1941. From the German *Blitzkrieg,* or "lightning war." Also used as a verb meaning to destroy in an air raid.

Buzz bomb See *V-1.*

D-Day The day on which British, Canadian, and American armies landed in Normandy, 6 June 1944.

Depth charge Large container like oil drum, filled with high explosive, dropped or fired from escort vessel. Explodes when preset depth is reached. Principal antisubmarine weapon.

Doodlebug See *V-1*.

Escort vessel Although battleships, aircraft carriers, and cruisers sometimes escorted merchant ships, the term usually applied to smaller vessels specially equipped for antisubmarine action, such as destroyers, frigates, corvettes, and converted mine-sweepers.

Flying bomb See *V-1*.

Giant German WWI multiengined very large bomber, the Siemens-Schukert R VIII.

Gotha German WWI heavy bomber.

Little Blitz German air raids on England conducted mostly in February and March 1944.

Radar Device using reflected radio waves to detect object in sky or on surface of sea.

U-boat German submarine, from *Unterseeboot*. Never applied to Allied submarines.

Zeppelin German WWI lighter-than-air dirigible, also called airship. Used as bomber.

Janie with survivors of our private V-1, still living in "temporary" billet in 1949 – "George the Second," "Billie," and Miss Hards. Photo by the author

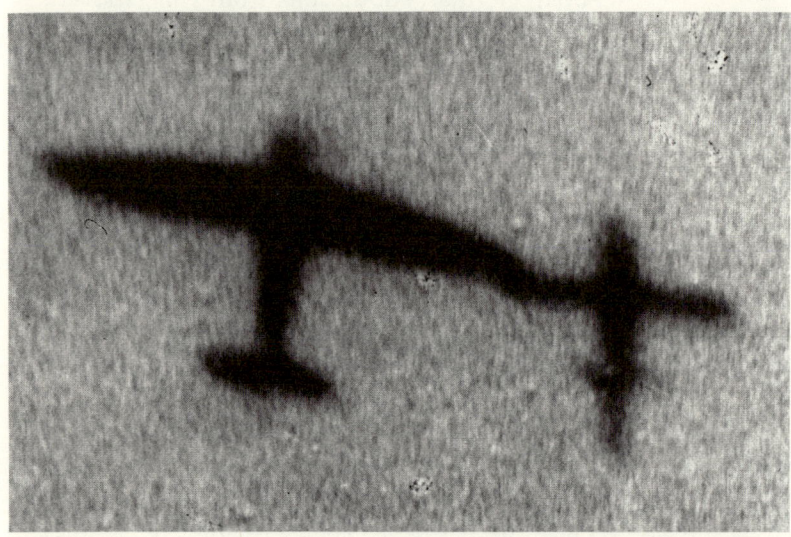

In daring manœuvre at 400 m.p.h., Spitfire pilot is about to flip smaller V-1 over so that it will crash in sparsely populated country, "Bomb Alley," 8 August 1944. Courtesy of the Trustees of the Imperial War Museum, London, #CH16281

Women rescue workers demonstrate how government-issued Anderson shelter survived V-1 blast that devastated houses. Courtesy of the Trustees of the Imperial War Museum, London, #HU635

Members of Women's Auxiliary Air Force handle an obstreperous barrage balloon. Courtesy of the Trustees of the Imperial War Museum, London, #CH21007

Seen from Fleet Street roof, V-1 hurtles down towards Drury Lane, Central London. Courtesy of the Trustees of the Imperial War Museum, London, #HU636

Photo taken three minutes after V-1 had exploded on impact at Clapham Junction shows emergency services already at work, June 1944. Courtesy of the Trustees of the Imperial War Museum, London, #HU36166

Vast area of destruction north-west of St Paul's, the scene of the Second Great Fire of London, 29 December 1940 – still a desert in 1949. Photo by the author

Women gunners in the Auxiliary Territorial Service assist artillerymen in firing coastal anti-aircraft gun at incoming V-1s. Courtesy of the Trustees of the Imperial War Museum, London, #H14986

Extraordinary time exposure shows white line where V-1 flew from left to right until blown up by anti-aircraft guns on lower right. Trail was made by flame from ram-jet engine. Specks along trail are shell bursts, some of which jolted target. Courtesy of the Trustees of the Imperial War Museum, London, #HU2646

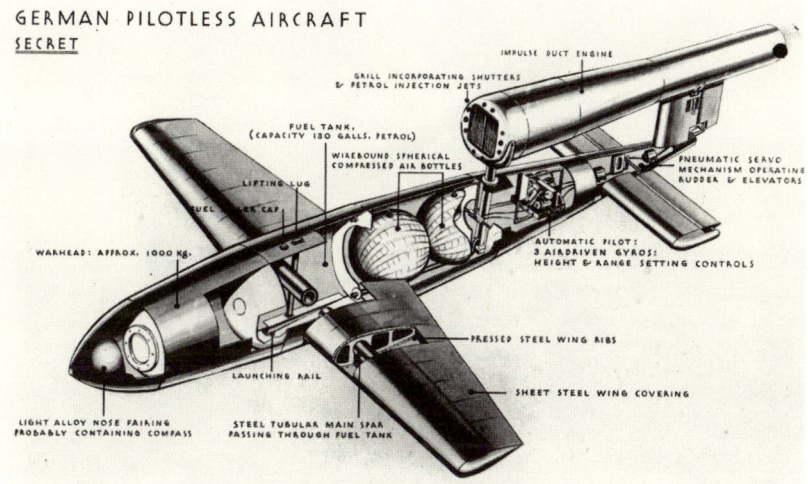

GERMAN PILOTLESS AIRCRAFT
SECRET

Carrying a ton of high explosives, the V-1 pilotless aircraft (flying bomb, doodle bug, buzz bomb) flew about 360 m.p.h. at an altitude of 2,000–3,000 feet. Launched from many sites in northern France and later from aircraft, they were aimed at London but had no specific targets. At the height of the attack they were destroying 20,000 houses a day. Courtesy of the Trustees of the Imperial War Museum, London, #CL4431

Four-stacker destroyer HMCS *St Croix* was sunk in mid-Atlantic by new secret weapon, the acoustic torpedo. Eighty-one crew members were rescued by HMS *Itchen*, all but one of whom were drowned two days later when *Itchen* herself became another victim of the new menace. Courtesy Canadian War Museum, Ottawa, #NMC75–12038

Frigid gales and spray made convoy operations even more dangerous off Canada's east coast. Exposed equipment and the entire superstructure became sheathed in tons of ice, adding the imminent risk of capsizing to the crew's worries. Here a Canadian escort vessel guards a medium-sized freighter in a rough sea. Courtesy Department of National Defence

Travelling 4,000 m.p.h. at an average altitude of 55 miles, this monstruous V-2 rocket carried one ton of explosive from Holland to London in four minutes, arriving ahead of its sound. Note man on high platform.
Courtesy of the Trustees of the Imperial War Museum, London, #CL3407

660 lb. Zeppelin bomb in 1915 turns Bartholomew Close into a grim anticipation of WWII, 8 September 1915. Courtesy of the Trustees of the Imperial War Museum, London, #LC56

CHAPTER ONE

London on D-Day

The English Channel was filled with the greatest fleet of ships the world had ever seen. The date was Monday, 5 June 1944. But on the Isle of Dogs I was completely alone.

The Isle of Dogs is not an island; nor were there any dogs to be seen that quiet summer evening. Situated about six miles east of central London, it is surrounded on three sides by water as the river Thames makes a U-turn. King Charles the Second kept spaniels there. But now there were neither spaniels nor people.

At its tip was a tiny green park. The rest of the little peninsula was nothing but roofless houses, empty basements, ruined walls, and desolation. Hitler's bombers had caused this destruction four years earlier.

Directly across the river stood the noble buildings of the Royal Naval College, Greenwich, bathed in the soft orange glow of the setting sun. Designed by Christopher Wren, they were built on grounds formerly occupied by a medieval palace where King Henry the Eighth had lived with his queen, Anne Boleyn. There she bore him the infant daughter who was to become Elizabeth the First. Anne was taken from here upstream to be beheaded in the Tower.

Beyond the college was a lovely park, in the centre of which, on the crest of a hill, sat the famous Royal Observatory astride the zero meridian, represented by a groove in the pavement outside.

Lying at the foot of the observatory was the beautiful white seventeenth-century Queen's House. In its grounds Sir Walter Raleigh spread his cloak across a puddle so that Elizabeth I would not muddy her royal shoes. In the far corner of the park was a deer enclosure. You will hear more about these poor deer later.

The only evidence that we were at war was the balloon barrage, silver blobs dotted here and there against the sky, patiently protecting us from low-flying bombers.

I descended in the ancient lift beside the river to the deep, long pedestrian tunnel under the Thames by which I had come. It was dank, dimly lit, and totally empty of other people. Water dripped down in many places. An ocean-going freighter that was approaching upstream when I descended must have passed over my head. I was glad when the lift on the other side disgorged me safely above sea level.

I strolled along King William's Walk, a street between the college and a row of houses that were soon to be the scene of a horrible tragedy.

Invasion was in the air everywhere. For weeks no one had been allowed to visit the southern portion of England. The country was bulging not only with British and Empire troops but also with a vast number of Americans, as well as many gallant men and women who had escaped from the Continent to fight their oppressors from this island. Everyone was speculating as to where the invasion would take place. It all seemed to me like excellent camouflage for a noninvasion. I had been nurtured in the values of air power and, since my arrival in Great Britain three weeks earlier, had seen something of its awesome effects. We knew that we were pounding Germany from the air around the clock, the Americans in the daytime and the British and Commonwealth forces at night. It seemed the height of idiocy to go to all the trouble of landing an army on a heavily defended coast when all we needed to do was keep up the bombing for another winter or two and Germany would most certainly surrender. Surely we weren't going to send soldiers back, fighting inch by inch, through a land dotted with the graves of their fathers and uncles, past such horror-filled, names as the Somme, Vimy, and Ypres. However, the combined chiefs of staff had not consulted me and foolishly went ahead on their own.

I was a meteorological officer (met. o.) attached to the Royal Navy, on loan from the Royal Canadian Navy, and was taking an intensive course in nautical meteorology at the Royal Naval College. Since the dormitories were full, many of us were required to hunt in the neighbourhood for our own accommodation. I had found a pleasant room in a house at 42 Ashburnham Place, about ten minutes' walk from the college. My attic room had a dormer window looking south, a fact that had no significance at this point but would later give me a good view of history being made. I walked home and went to bed, oblivious of the momentous events shaping the destiny of the world a few dozen miles to the south-west.

I dozed off but did not sleep well that night. For one thing, the wind was gusty and it kept rattling my window, waking me up every so often. I was also disturbed by even more aircraft than usual thundering overhead towards the Continent. The sounds infiltrated my dreams, and I dreamt that the invasion was taking place all around me. When I woke up next morning I said to myself, "It won't be today, it's too windy."

I went to breakfast where we ate all our meals, in the Painted Hall of the Royal Naval College. The lofty ceiling was covered with classical allegorical paintings in the manner of the seventeenth and eighteenth centuries. Along the walls there appeared to be niches containing life-sized statues, but on close examination these were found to be optical illusions, two-dimensional feats of *trompe l'oeil*.

We were served on long oak tables that extended the length of the room to a raised dais at the far end, where the officers of the college sat. It was there that Nelson's body lay in state after the Battle of Trafalgar. Silver candelabras were set at intervals along the tables, adding a most unseamanlike touch. Entering this magnificent setting under a huge dome always filled one with awe and respect and seemed to bolster the young officer's desire to live up to the traditions that surrounded him.

We shared this hall with one or two hundred other British naval officers taking courses at the college, ranking from midshipmen to distinguished heroes from the three services studying combined operations. Those of us in meteorology sat together along an assigned table. Our group consisted of about fifteen British officers and four Canadians, as well as half a dozen officers of the Women's Royal Naval Service (WRNS), known universally as the Wrens for obvious reasons.

Breakfast was as usual. I mentioned that I had had a restless sleep and dreamt that the invasion was taking place. One or two of my fellow officers ribbed me, saying, "You're a fine met. o., dreaming up an invasion on such a windy night."

At 9:00 we resumed our course. Several classmates who lived at home had heard the BBC quote German radio to the effect that parachutists had landed in northern France north of the Seine. This was interesting, but not unusual. As there had been no confirmation we did not consider it significant. They could have been Allied bomber crews bailing out, commandos, or dummies intended to make the defenders nervous. We continued working on our weather maps.

My contemporary account describes the rest of the day as follows:

After lunch I was strolling across the courtyard in the sunshine when a midshipman in our course came up and said, "Today's D-Day! The invasion started this morning." He had just heard it on the gunroom wireless.

Needless to say, this was a great moment, one for which we had waited ever since 1940. I mentally threw my cap in the air and turned a few handsprings, but at the Royal Naval College this isn't done for just an invasion, so physically I didn't do anything but thank him for telling me and ask for more details. He quoted the figures of the large numbers of ships and aircraft engaged and said that things were going well.

We parted and I found several other people and told them. There was no great panic. If the fellow didn't know already he was sceptical and tried to pretend he suspected it all along; if he had just heard he was superior to the point of being rude. I bit a subbie's head off for telling me the news one minute after the midshipman had.

When we reassembled after lunch, I felt we should sing the National Anthem, or something, but nobody else seemed to, so we just started work again. A few of us said, "Well, well, so this is D-Day," from time to time but soon got discouraged by the unappreciative audience.

Everyone, of course, was very happy. The fact that we got so cranky showed how much it really meant. But people who had undergone what the English had were incapable of getting enthusiastic over D-Day. It was an important turning point in the war, but it seemed to be looked upon more as a follow-up to El Alamein, Tunisia, Sicily, Salerno, and the air bombardment of Germany. They realized from bitter experience that this was only another step and that much more had to be done. Moreover, a great many people knew that their loved ones were in the battle, and that added a note of strain that sobered the general atmosphere.

(One of our Wren officers, Gwen Kimber, knew that her husband, a naval officer, was assigned to clearing the underwater-explosive obstacles ahead of the landing craft. In due course we were happy to find that he had survived his perilous task in good shape.)

My account continues:

Later, letters from Canada and the US gave us the impression that the general rejoicing and excitement on D-Day were much greater than anything I saw.

At tea time that evening papers were seized with gusto, and each was read by a group of several people at once. Everyone seemed very pleased with the progress reported, but there was no wild tumult. I have seen many a promotion cause more turmoil in a wardroom, even at tea time.

In the evening I went over to a rooming-house where my three fellow Canadians lived, and we listened to the radio. The BBC's handling of the nine o'clock news was vivid without being sensational, and it was obvious from Prime Minister Winston Churchill's report that things were going remarkably well. The king's speech was moving, and the eye-witness accounts brought home to us very keenly what so many men had gone through that day on our behalf.

I admired what I considered General Dwight Eisenhower's* brilliant deception in implying that this was the main thrust. It had always seemed to me, armchair military strategist that I was, that if we were ever going to invade Europe, the sensible place would be where we had left it – that is, near Dunkirk or Calais. It seemed absurd to land so far west that we would have to go through three hundred miles of heavily defended countryside before reaching Germany. Hitler had the same erroneous opinion. We now know that he, too, thought that the main thrust would come across at the nearest point, and that the Normandy landings were but a diversion. He kept his main armour in that area for several days, largely due to the tremendous deception that had been so carefully built up over a period of months – canvas tanks galore, dummy aerodromes, fake wireless traffic in the southeast

---

* Eisenhower had been appointed supreme commander of the Allied Expeditionary Forces in December 1943.

of England, the brilliant idea of making it appear that General Patton was commanding a huge army in the southeast.

Virtually all German spies in England had been caught and either executed or persuaded to work for us – sending out false reports thought up by the highly imaginative British Intelligence.

However, we had one concern. Hitler, Goebbels, and their propagandists had been claiming for months that they had designed three vengeance weapons* that would win the war. We knew that they were liars and boasters who had issued a number of false claims in the past – they declared on several occasions that they had sunk the aircraft carrier *Ark Royal*. Nevertheless, there was always the disquieting possibility that they were telling the truth. No secret weapons had been used against the landing forces in Normandy thus far, and the tone of German propaganda implied that such weapons would be directed against the civilian population of England. We had worried about gas attacks ever since World War I. Everywhere one went in Britain, there were decontamination centres for cleansing people spattered by mustard gas, and everybody in Britain had been issued a gas mask. For the first year or two of the war, the British had been required to carry them at all times and were fined if they were found without. I remember a cartoon in *Punch* early in the war in the series called "The Changing Face of Britain." It consisted of two panels, both depicting the Lost and Found office at a railway station. The first was full of everything imaginable – suitcases, umbrellas, parrot cages, potted palms. In the second was nothing but gas masks from floor to ceiling.

We had all been given gas drill in our early days in the armed forces, but now merely had to keep our gas masks with us when travelling from one base to another. We could leave them in our rooms as long as we were never without them overnight. Before going to bed, I had a look at my gas mask to be sure it was in good order and gave myself a refresher drill in putting it on quickly with my eyes closed while holding my breath. Having thus prepared for Total War, I went to sleep very much aware of what so many members of the Allied armies were enduring on the beaches of Normandy. The night was uneventful and on awaking I was happy to find my concerns had been unwarranted – so far at any rate.

---

* Fortunately, V-3s were never developed.

CHAPTER TWO

London, the Old Warrior

Before describing the attacks of the V-weapons, it would be well to review London's past experience in air raids. Many people do not realize the extent to which England suffered from aerial bombardment in World War I.

The sixth of August, 1914, is a terrible date in the history of our so-called civilization. The Germans opened up a new form of terror – the aerial attack on civilians. A Zeppelin dropped thirteen bombs on Liège, Belgium, killing nine civilians. Thus, Liège is at the top of the sombre scroll that includes London, Guernica, Warsaw, Rotterdam, Liverpool, Coventry, Cologne, Hamburg, Berlin, Dresden, Hiroshima, Hanoi, the Falkland Islands, Tripoli, Baghdad, and Tel Aviv. The world was aghast at the attack on civilians, and as more raids took place on cities the word German became synonymous with "barbarian"; they were called "Huns" after the tribe that devastated Europe in the fifth century.

The first attack on Britain was made on 21 December 1914 when a lone aircraft raided Dover, one of a great many such forays to come. Ramsgate alone endured 119 air raids. The damage from these minor stings was never widespread, but the sense of outrage and the fear of future attacks were pervasive.

During the next four grim years England was never long free from danger. The first substantial attacks were made by Zeppelins, commencing on 15 January 1915, when two civilians were killed in Yarmouth and a woman and child in King's Lynn.

These strange ships of the sky were about as long as two football fields and were powered by several engines. By the end of the war they could attack from about 20,000 feet, well beyond the range of guns or fighter aircraft. They carried several thousand pounds of bombs, both high explosive and incendiary. When flying at great heights above the clouds they could not be heard from the ground, and the first indication of their presence was the crashing of the bombs. Often several airships attacked together; on one occasion fourteen of them flew simultaneously over the metropolis.

On 17 August 1915, bombs fell on the north-eastern suburbs of Leyton and Wanstead Flats, killing ten, injuring forty-eight, and destroying the railway station, a tram barn, and many houses. Another serious incident took place on 8 September. A 660-pound bomb, at that point the largest ever used, fell on Bartholomew Close in Central London, causing the city's first air-raid conflagration, eventually doused by twenty-two fire engines. These and other similar incidents were comparable to many of those in World War II.

There were sporadic attacks by groups of Zeppelins until 6 August 1918. However, these unwieldy gas bags suffered severe losses due to strong winds and attacks by fighters. They were gradually superceded by aircraft, particularly Gothas. On 13 June 1917, twenty of these new bombers took off for London. Fourteen reached the capital around noon and bombed at will as the fighters had not been able to locate them on the way. Seventy-two bombs fell near Liverpool Street Station, and one hit a school in Poplar, across the Thames from Greenwich. Eighteen children were killed and many others horribly mutilated. All in all, the raid killed 162 civilians and seriously injured 432.

Eye-witness accounts of the raid sound like something from World War II. All the elements are there – mangled bodies, the moaning of the injured, piles of rubble, flying glass, volunteer rescue workers, and choking dust. In 1935 the *Evening News* printed two letters from survivors of WWI air raids. The first concerned a Zeppelin attack, the second the Gotha raid that destroyed the school at Poplar. Both occurred in East London a short distance from where, exactly twenty-seven years later, the first flying bomb demolished several houses, including the home of Joseph V. Waters, to whom I am indebted for these clippings (see chapter 24, "1989" and "1990").

On hospital leave from France, I was proceeding home in Bow on the evening of September 23, 1916. The warning was circulated that Zepps were making their way to London.

Reaching home, I warned my wife and family to take cover and, placing my kit-bag in the hall, went out into the garden. The Zepp was moving in direct line with our house and was held by the searchlights, with shells bursting underneath her envelope. As she came on, I went in and joined my wife and two small daughters, who were very frightened.

Suddenly, with an awful roar which seemed to rock our house on its foundations, a bomb fell. Our front door burst open, the lock being wrenched off, and the front (iron) entrance-gates were blown off and broken. Every window in the front was blown out.

After making sure my family were safe, I went to Bow-road, a few yards away, where a large public-house, the Black Swan, had been wrecked. Only the carcass of it was left standing, and a heavy pall of black dust hung over the ruins.

I and others groped our way amongst the debris, searching for any victims who might be alive. Lifting some flooring, we discovered the wife of the licensee, Mrs Reynolds, lying in the cellar, where she had been blown by the bomb. It had struck the house dead in the middle, taking all the floors to the basement.

Firemen found a baby stuck in the rafters. Fortunately there had been no customers in the house at the time.

The Zepp then passed over our house, dropping another bomb on cottages a few yards away, and another on a timber-yard, causing a big fire. I spent the remainder of my leave making good the damage to my house.

From W.G. Roberts (Lieut., London Regiment),
Valetta, Cairo-avenue, Peacehaven, Sussex.

The newspaper editor adds the following information:

[The Zeppelin that bombed Bow-road on September 23, 1916, was the L33. She was picked up by searchlights and badly holed by gunfire. Pursued in an aeroplane by Lieutenant Brandon, she came down intact at Wigborough, near Colchester. The bomb at the Black Swan killed the proprietor's daughters, Miss Cissie Reynolds and Mrs Adams, his baby grand-daughter, a 63-year-old woman, Mrs Potter, and a neighbour's young son. The same night two other airships bombed Streatham,

Brixton, Kennington, Hackney and Bromley. London casualties: killed, 23; injured, 66.]

Another letter follows.

Although I am one of the "new generation" myself, I have imprinted indelibly in my memory an experience of horror.

On the morning of June 13, 1917, I was sitting as usual in my classroom in the infants' department of North-street School, Poplar. Our teachers had been warned of an approaching air raid and were endeavouring to keep us calm by getting us to sing all together. Soon, however, the noise of the anti-aircraft guns and the detonations of the enemy's bombs became audible above even our shrill voices.

I am never likely to forget that scene, because the next second a bomb, passing through the upper floors of the school, exploded beside me.

The official announcement of 18 children killed (average age six years) and many seriously injured, does little justice to the terror of that moment. The eleven other children in my row of twelve were all killed, while I myself, although I managed to run home somehow, was unconscious for six weeks, eventually to make a slow recovery.

Sometimes I pass a local memorial stone which has on it: "IN THE MEMORY OF 18 CHILDREN OF NORTH-STREET SCHOOL, KILLED BY AN ENEMY BOMB ON JUNE 13th, 1917", and I think "Never Again!"

From Miss I.A. Major, 45 Dee-street, Poplar, E.14.

The editor adds: "[This was an incident of the daylight raid by 14 German aeroplanes: nine boys and seven girls (mainly in the infants' room) were killed, and of 31 other injured two died later; all the raiders escaped.]"

At various times bombs fell on such familiar landmarks as Waterloo and London Bridge stations and Green Park. Another landed in the Dean's Yard at Westminster Abbey but failed to explode. On a warm and pleasant Sunday afternoon, 12 August 1917, the seaside resort of Southend was bombed indiscriminately by Gothas; thirty-two were killed, most of them women and children who had come from London for a day of fun. Let no man or woman who did not experience these raids criticize the severity of the Treaty of Versailles and the postwar treatment of Germany.

Despite the horrors, there were indications that the traditional spunk of Londoners was undiminished. Thousands of people went into the parks at night during raids "to see the Hun air show."

Before the end of the war the Germans had built a new bomber, enormous by any standards, the Siemens-Schukert R-VIII, or Giant. Powered by six engines, with a wingspan of approximately 150 feet, it was larger than any aircraft used in Europe during World War II. Its wingspan was only twenty-two feet shorter than that of the much heralded B-2 "Stealth" bomber, which first flew on 17 July 1989. Production difficulties kept the supply limited, but on nineteen occasions a handful of these Giants raided England, mostly in 1918.

One of Hitler's many mistakes was his failure to revive the development of the multiengined heavy bomber. Throughout the war he never had such a weapon. The aircraft of the Blitz were designed primarily for work with the army. London could be thankful for that miscalculation.

During wwii London could also be thankful for the experience gained in the attacks of the earlier conflict. The whole paraphernalia of air defences used in wwii had been established by 1918 – air-raid shelters, a balloon barrage, anti-aircraft guns, searchlights, the blackout, sirens, and the network of observers. The differences were in the scale of the attacks and the improvement in the hardware.

According to Raymond Fredette, the total number of raids by Zeppelins during the war was fifty-one. However, *The Encyclopedia of World History* states that there were sixty Zeppelin raids in 1915 and 1916 alone. Fredette's figure for raids by airplanes is fifty-two. This presumably does not include sneak attacks along the coast, as the *Blue Guide* gives 119 for Ramsgate alone. At any rate, there were far more raids than the English would have liked.

The total number of casualties in Britain from air raids during World War I was 4,830, of whom 1,414 were killed. These figures were minute when compared with casualties in the trenches and in that strip of hell known as No Man's Land. The damage to factories and homes was also of minor importance except to those who were directly involved. But undoubtedly the raids made life miserable for many millions of children, women, and men for a period of four years and contributed greatly to the hatred of the Germans.

Thus, when World War II started every Londoner over thirty had vivid recollections of many air raids and had developed a number of personal survival techniques, such as wearing woolly socks in the shelters or working on crossword puzzles. Thousands of people remembered their favourite places of safety and returned there when the raids started. One such spot was on a road that ran underneath the tracks at London Bridge railway station. The authorities had great difficulty convincing the poor souls who sheltered there that it was no longer safe from the new bombs.

Britain's Prime Minister Neville Chamberlain declared war in a radio speech on the 3 September 1939, two days after Hitler had invaded Poland and had failed to withdraw his troops in accordance with the demands of Britain and France. As soon as his broadcast was over, the terrible sound of air-raid sirens was heard. Everyone assumed that this foretold the arrival of the great air armada that had been predicted by all military authorities for a decade. People trooped into the air-raid shelters that had been hastily constructed during the year's respite given by the Munich Agreement. Many persons described how they turned to have one last look at their house before it was destroyed. They all admitted to having been frightened. As the nation crouched, fearing the worst, the All Clear sounded and everyone came to the surface feeling rather foolish. One lone friendly aircraft had crossed the coast and was mistaken for the enemy.

This alert had a curious effect on the British psychology. They felt that they had overreacted, and thus, when the real raids started, they were determined not to get the jitters.

There was virtually no enemy air activity over England for the first few months of the war and the British merely dropped leaflets over Germany. But soon after the fall of France the Germans started attacking in the daytime. The date of the first major onslaught on the island itself is generally considered to be 12 August 1940, marking the start of what became known as the Battle of Britain. First they mainly attacked the radar stations on the coast, then the fighter aerodromes, but the pressure was not sustained enough to be crippling. The Germans were doing a great deal of damage but gave up a little too soon in each case.

On Saturday, 7 September, a huge daylight raid took place on the dock area of East London. A tremendous fire resulted and there was massive damage. While this was not new to Londoners, it was much more devastating than the attacks of the First World War.

On 13 September a Heinkel III bomber deliberately hit Buckingham Palace in daylight, causing considerable damage. This was an enormous mistake on the part of the Germans as it united the population from top to bottom. The queen was quoted as saying that she could now look the East Enders in the face.

After some weeks of daylight raids, the Germans were suffering such heavy losses from Fighter Command that they had to give them up. The last full-scale daylight operations against London were mounted on 30 September concluding the Battle of Britain. From then on the heavy raids were at night. Thus the Blitz began.

Based on the German word *Blitzkrieg*, or "lightning war," "blitz" became both a noun and a verb and passed into the English language forever. The term generally referred to the night raids that started in October and continued until May 1941. There were heavy attacks for many consecutive nights – nearly two months on London – and then attacks on other cities for several days each. Frequently the bombers returned to London, especially the night of 29–30 December, when the second Great Fire raced through Central London, wiping out a vast area of business premises north-east of St Paul's Cathedral. The general morale did not break and war production continued with relatively little interruption. Finally in May 1941, Hitler withdrew his forces to attack Russia, much to everyone's relief except the Russians'.

Forty thousand people were killed in the raids on Britain up to this point, and 46,000 were injured. The official Royal Air Force (RAF) history states that 600,000 men were engaged in civil defence. This presumably means "people," as many women served with distinction and gallantry.

In the next two years there were enemy attacks, but they were mostly swift forays of a few aircraft aimed at specific targets. They had decided in Germany that if bombing the cities did not destroy the British will to fight, bombing their famous buildings would. Attacks were made on a number of places that were described in guidebooks such as the well-known Baedeker; hence these were called the Baedeker raids. They often took place in daylight. Canterbury Cathedral was damaged, as were other historic buildings. In one of those raids there was an attack on the Royal Naval College by a single raider who flew up the Thames and dropped a bomb on the quarters of the second in command, killing the commander during his lunch hour. This was at the far end of the building in which we took our classes. Severe damage was still visible.

In February and March 1944, there was another series of night raids of a particularly unpleasant nature. This was called the Little Blitz. It was noisier than its notorious predecessor. More anti-aircraft guns had been added to the defences as well as new rocket-firing weapons that made a tremendous racket. A policeman told me that when a rocket passed overhead it sounded like several railway locomotives roaring by at once. The noise got on everyone's nerves.

At the same time the Germans also dropped antipersonnel bombs of a nasty nature, soon named "butterfly bombs" because of their appearance (which resembled a pair of handcuffs rather than a butterfly). They would catch on wires or twigs and glitter attractively. Curious children would touch them and have their hands blown off. These diabolical devices were of no military value whatsoever. They were strictly designed to kill or maim civilians.

In the Little Blitz bombs had damaged the women's residence of the Royal Naval College. This was just across the street from the main buildings. Damage was still being repaired when I arrived. In the raid a Canadian Wren met. officer named Helen Partridge was slightly wounded in the wrist by flying glass. This incident proved that Canadian women in the armed forces were subject to the same risks as men, although receiving lower pay and fewer benefits. It highlighted the fact that they too were in the front line.

When I arrived in mid-May all was peaceful. But I found a disquieting undercurrent of fear. On one occasion I shared a railway carriage with a middle-aged man who talked about the Little Blitz. He looked very frightened as he described how he had had to throw himself on the ground when a bomb landed nearby. Others said how unnerved they had been at the renewed attacks and described horrible incidents caused by the butterfly bombs.

I also heard that when the king and queen visited a bomb site, as they had done so often during the Blitz, they were booed – something that had never happened during the earlier stages of the war. All this made me quite uncomfortable, and I was unsure how the home morale would stand up to the advent of Hitler's much-boasted-about Vengeance weapons. Time proved that I needn't have worried.

CHAPTER THREE

Liverpool, Another Veteran

My first signs of air war were when our convoy approached Liverpool on 14 May 1944. I had just crossed the Atlantic for the first time as the lone passenger in the merchant aircraft carrier *Empire Macoll*. She carried a full load of wheat and four old aircraft, which operated from a flight deck that ran the length of the ship. Our convoy, HX289M, formed into a single line and steamed up the swept channel through our minefield. Concrete towers supporting anti-aircraft guns rose out of the water. A Spitfire streaked overhead. A flotilla of escort vessels hurried home for a rest.

The buildings of Liverpool became clearer. In the channel ahead I saw what I thought were two spar buoys – long thin poles extending several feet above the water. As we drew closer, the grim truth hit me. This was a sunken ship, doubtless the victim of air attack several years earlier. Our wash swept over her and splashed against the tops of the masts, the only tombstones for some brave sailors.

We entered the narrow harbour that is the mouth of the Mersey River and anchored. Our sister ships edged past us slowly to the berths assigned them for the night.

The industrial Liverpool shore showed no signs of damage. It was Sunday evening, and on the opposite side an amusement park was in full swing. The trees were covered with fresh leaves, a welcome sight. We had not seen any green in Halifax since the previous autumn, and after sixteen days of grey waves, grey ships, and grey skies, the colour was most refreshing. Occasionally a ferry passed near us, bringing couples back from a pleasant afternoon at the park. The men were almost all in uniform and the women wore brightly coloured summer dresses. They waved to us as they passed. Apart from the uniforms the total scene was one of peace, with no sign of the terrible war the country was waging.

Dusk came and I turned in. I had grown accustomed over two weeks to throbbing engines and constant rolling, and now the deathly silence kept waking me up. But I was excited at the prospect of seeing England and could hardly wait for morning.

Very early the tugboats arrived and nudged us into the maze of docks that are characteristic of many European seaports, with their large tides and shallow harbours. The tide was now right, and we slowly worked our way through canals, past lock gates, swing bridges, railway tracks, cranes, ships, and warehouses until we came to our dock. The minute our gangway touched land, dozens of workmen swarmed aboard and began making minor repairs, removing our hatch-covers and connecting the flexible tubes that were to suck out our grain. Within five minutes the unloading process was fully underway. I was naively amazed at the Old World's speed and efficiency.

I phoned for a taxi and while waiting looked around. Several times during the crossing my shipmates had said, "You won't see any damage now. It's all been cleared up and repaired." That certainly seemed to be the case. Then I noticed a nearby pile of twisted girders. My stomach suddenly tightened as I realized that I was standing where there had once been a warehouse. I was at last in actual physical contact with the war.

The taxi drove into town past other signs of bombing – here and there an empty building with no roof, in other places basements with weeds growing through cracks in the floor. I talked to the taxi driver about the Blitz. He was a little man in his forties.

"Oh yes, we had it bad all right. It was tough then. Didn't get much sleep, driving all day. I'll never forget one night," he went on. "We'd had it for several nights in a row. I was pretty tired. I had a few beers at the local and went to bed early. The bombers came over soon, and about an hour later a parachute land mine came down in the next street. Fairly lifted me out of bed. Blew all the windows in, and a beam fell across the floor. Fellow from the next room helped me get out. I was good and sober by this time.

Then we pulled on a few clothes and went out and spent the rest of the night digging for people."

"Find anyone?" I asked.

"Yes. A woman with a baby in her arms."

"Dead?"

"Been that way for hours."

Feeling faintly nauseous and wobbly-kneed, I had him drop me off at Lime Street Railway Station. This was my first acquaintance with one of these Victorian gothic eyesores, dirty and ornate on the outside and dirty, dark, cold, and smokey inside. Having found that the checkroom was called the Left Luggage, I set out to explore the town. Some massive office buildings were empty shells, others were being repaired, while many were still intact. I was shocked when, on rounding a corner, I came on an area of several blocks that was totally empty – nothing but basements and the odd wall. This devastation, overseen by a statue of Queen Victoria, staggered me. I said to a nearby elderly gentleman, "It must have been terrible."

"Yes," he replied. "I remember passing here one morning and seeing a car lying on its side beside the statue. The district was a shambles."

In due course I presented myself at the headquarters of the commander in chief, Western Approaches, the nerve centre of all North Atlantic convoys. I was shown around the Operations Room and the met. office.

That night I took a sleeping car to Glasgow. First-class bedrooms on the trains were reserved usually for VIPs or the wounded, so I booked a third-class berth, there being no second class in Britain for reasons unknown. Soon afterwards I wrote the following account. It may seem a trifle exaggerated except to those who have spent a night in a wartime third-class sleeping compartment.

Four people bunk in one cabin, on hard narrow slabs two deep, running athwartships instead of fore and aft as at home. One blanket is assigned per slab and there are no sheets, but a section of what felt like a thick plank covered by a pillow-case is at one end.

The four of us took turns undressing on the small floor space, while the rest lay on our slabs out of the way. In due course we were ready for bed and rolled ourselves up in our blankets. Mine was an upper berth, a terrific height off the floor. I was afraid to look over the edge.

As soon as the train got moving the cabin cooled off. First I put on a sweater, then my trousers. I turned over once too often in one direction without reversing in between and found the blanket had been drawn underneath me as through a roller and I was completely uncovered. I rewound myself and tried to get to sleep ...

I continued to turn over and over, as the slab paralyzed whatever side I was lying on after a few minutes. Each time I had to rewind my shroud and redrape my coat. Finally I fell asleep. About three times in the night I woke up from dreams of icebergs to find my coat had fallen down onto the floor. I had to swing down, grope for it in the dark, and climb back without stepping on anybody on the way up. It was hazardous, but actually not as much as [it would have been in a North American Pullman car], as once you gain the correct altitude you don't have to hover while you find the gap in the curtain.

In due course morning arrived, and after a towelless, soapless, cold-water wash, I dressed in what were mostly my own clothes, and we reached Glasgow.

This account appears to have been expurgated in the interest of morality. My distinct recollection in later years was that when I entered the compartment one of my fellow passengers was wrapped in a blanket, sound asleep. When the rest of us awoke, the same passenger was dressed and ready to leave – a Wren officer, sitting demurely on the edge of her bunk.

CHAPTER FOUR

London's Old Wounds

I spent two nights at the Canadian naval base at Greenock, near Glasgow, and an unforgettable day in lovely undamaged Edinburgh, with which I fell madly in love. Then, a night train to London, sleeping in a day car.

The train arrived in London at an early hour, but the sun was already shining brightly. I was to report to the Admiralty, the headquarters of the British navy. There was no point in hurrying, as I had been told that little happened there before 10:00 A.M.: it was considered ungentlemanly to start work earlier. They made up for it with prodigious efforts after teatime.

I checked my suitcase and started walking from King's Cross Station towards the central area. There was little evidence of damage, but everything looked dingy. Paint had been unavailable for civilian use for years, and the soot of ages had turned the metropolis grey and black. On the other hand, evidence of war was abundant. This was a city that had obviously given great thought to survival.

Many basement or ground-floor windows were bricked up to strengthen the lower floors, often used as shelters by the residents. Large plate-glass shop windows had been replaced with plywood, solid except for a small glass area in the centre through which merchandise could be displayed. Other windows in buildings and buses were taped in lattice patterns to reduce the risk of flying glass.

Everything that one might bump into during the blackout was painted white – tree trunks, lamp-posts, steps, and so on, and the edges of curious structures called "blast baffles." These were short brick walls, six or eight feet high, extending out a foot or two from one side of a doorway and then turning at right angles to shield the entrance of the building from bomb blast, which could be devastating. The layers of whitewash on these obstacles were an indication of how many people must have bumped into them since they were first built.

Occasional signs pointed to public air-raid shelters. Sometimes these were built in the middle of the road, others were underground. Signs pointed to decontamination centres, a precaution against the dreaded mustard gas that fortunately never arrived.

Every few blocks an open space had been converted into an emergency water reservoir, the walls extending three or four feet above ground, open to the sky, and filled with water of considerable antiquity. Iron pipes ran from these along the sidewalks to supplement the supply from regular fire hydrants, providing splendid opportunities to break one's leg in the blackout.

Virtually all the public clocks had stopped running except Big Ben. Either the jolting effect of bombs and anti-aircraft guns or the shortage of skilled technicians had brought them to a grinding halt. Wisely, they had been set to read twelve o'clock as a reminder that they were not telling the truth.

After walking for some minutes I came to a small school, probably built in the nineteenth century, beside which was a paved playground complete with basketball hoop. The school had been badly damaged. Working on some equipment were figures clad in blue grey. A second look indicated that their hair was too long for airmen, and my ears soon told me they were sopranos. Extending upwards from a winch was a stout cable. My gaze ran up it to a great height where a silver barrage balloon shaped like a blimp hung in the air, bright against the pale blue sky. It was one of dozens scattered throughout London to deter low-flying attacks. I wondered whether the ground crew, members of the Women's Auxiliary Air Force, had been there when the school was hit. A nearby tent was their home around the clock and in all weather.

Soon I was walking around the gradually awakening centre of London. I had an old guidebook and a large map and was thrilled to see so many famous sights in an hour or two of walking. Bomb damage, however, was visible in many places, but not utter devastation along my route. Warehouses beside the Thames were blackened and empty, their walls still standing but their insides burnt out. In the north transept of St Paul's Cathedral, a huge hole showed where a heavy bomb had crashed through to the crypt. Some office buildings and churches were gutted completely; in others only the basement remained.

On the Embankment by the water's edge I discovered Cleopatra's Needle, the famous tall stone obelisk. Built in Egypt around 1500 BC, it had been tied to buoyant material and towed to England in the nineteenth century. It broke away in a storm and was lost for some time, but eventually was recovered and erected in its present place. Ironically, it bore several scars from Zeppelin raids in World War I. Danger from the skies was nothing new to Londoners.

In Whitehall, the wide street of government offices, several of the massive buildings were missing chunks of masonry from their façades. In a few places one could tell from the angle of the scars where a bomb had landed on the sidewalk, but the pavement itself was completely replaced and well weathered.

Between the Admiralty and Admiralty Arch lies the Whitehall Theatre. Undaunted by its august neighbours, it displayed a huge advertisement for the current show. Under the unanswered question, "Is it true what they say about Dixie?" was a vast coloured picture of the lady herself. She had obviously dressed hurriedly, having forgotten to pull her clothes up over her right nipple, which gazed out at the passing throngs in Trafalgar Square. Despite the popular belief that war is one long orgy, this was the only visual evidence I encountered in fourteen months to indicate that women are not the same shape as men.

I wandered into the little lane called Downing Street and saw Number 10 with a policeman standing in front. Today the public is not permitted to enter the street because of the risk of terrorist bombs, but at the height of World War II no such restriction applied and everything seemed quite peaceful.

After an awe-inspiring stroll around Westminster Abbey, I headed towards the nearby Houses of Parliament. On a grassy traffic island was a newsstand displaying piles of papers. The owner was standing behind the counter. As I approached, something about him struck me as odd. A few more steps and I saw the reason … the whole thing was a fake, another example of *trompe l'oeil*. Pointing out from a small opening in the concrete structure was a bona fide machine gun, placed there to defend the Mother of Parliaments. Whether it would have stopped the German army will never be known, but it undoubtedly gave comfort to the Honourable Members.

I continued along Whitehall to the Admiralty, which had finally realized that another day had arrived. I was signed on as a member of the crew of HMS *President*, the shore establishment to which all of us in the London area were appointed. In due course I went out to Greenwich to find a room.

CHAPTER FIVE

Greenwich and Its Civilian Fighters

The main street of Greenwich was lined with a mixture of intact buildings and ruins. The business section of the town was ugly to start with, and four years of bombing and a paint shortage had made it considerably worse. Every few yards were to be seen empty basements, walls propped up, sagging staircases, faded wallpaper exposed to the sky and blackened by fire or stained by sooty rain.

I rented my room at 42 Ashburnham Place. The house was half a duplex, although most of the neighbourhood consisted of row houses attached together, sometimes called terraces. They were all made of the usual East London grey yellow brick with grey slate roofs. Our house had a tiny front lawn and a larger garden at the back, separated from its neighbours by a brick wall about six feet high. This wall played an important part in our lives later.

One door away stood St Mark's, a grey stone Methodist church. It fronted on a main street, New Cross Road, running at right angles to Ashburnham Place. Beside it was another row of houses. Behind the first was an Anderson shelter, a curved corrugated steel roof partially sunk in the ground, capable of protecting three or four people reasonably well. This shelter also has a future role in the narrative.

Our house lacked such a shelter, but like its neighbours, had a downstairs that was half below ground with windows in wells. It contained the kitchen, a front bedroom, and a rear sitting room. The main floor included a dining-room and a never-used Victorian drawing-room, all prim and proper. The upper two floors each contained two bedrooms and a bathroom. My room was on the top floor with a sloping ceiling and, as mentioned before, a dormer window looking south over gardens and a chicken run. Beyond the gardens were the houses on the next street to the south. It was called Ashburnham Grove. The similarity of name to Ashburnham Place caused many mix-ups, one of them with heartbreaking consequences.

The landlady and owner of the house was Miss Edith Hards, a thin, short woman probably in her mid-sixties. She was tight-lipped, her pale face adorned with a pair of small wartime spectacles designed to fit inside a gas mask. She was never still for a second. Her gestures were frequent and jerky, and when she had nothing to gesture about she weaved constantly back and forth. My Scots-Irish surname was totally beyond her English capability. After getting the "Mack" part satisfactorily, the rest was simply a diminuendo of unintelligible syllables.

Her companion was a much older woman called Billie, apparently without surname. She was past eighty, white haired, red of face, tall and lean. I never found out whether she was a relative or just another boarder.

The permanent male residency consisted of an Irish longshoreman in his twenties named John, and a certain gentleman referred to at all times as George the Second. He seemed to be some sort of Arctic explorer as he was usually away at the North Pole. (In due course I found him to be retired, slender, white haired, blue eyed, and a regular at a nearby pub of that name. Some months later it was bombed, much to his annoyance, but is now restored and doing a roaring business.) No reference was ever made to George the First.

An even more mysterious inmate was apparently Miss Hards's grandfather. She quoted him occasionally, but only when referring to the time. I subsequently found him standing in the hall chiming the hour, twenty minutes early.

Our menagerie included a fuzzy dog whose fore was indistinguishable from his aft, and two cats that came and went as they pleased.

Miss Hards and Billie were the soul of hospitality, offering me cups of tea at odd times. I was the first Canadian the neighbourhood had ever seen. Although I looked like a British lieutenant, Royal Naval Volunteer Reserve (RNVR) Special Branch, I talked, in their view, like an American movie actor. Thus I was given the tender loving care usually bestowed on a curio.

The two ladies had suffered through a tough war, with shortages, rations, and queues as a constant background to the raids they had endured for three years in ww1 and four in this, with more to come. They had forgotten much of what little schooling they had received in the previous century, and their knowledge of things North American had largely been acquired in their rare visits to the "flicks." I doubt whether either could have found my continent on a map. I was startled once when Billie asked me out of a blue sky, "Do they have dogs and cats in Canada?"

After Miss Hards showed me to my room I went to wash up. This was my first acquaintance with English domestic plumbing of Victorian vintage. As I recorded at the time:

What I saw shook me to the core. The wash-basin was a china one on stilts. I had seen one like it once years ago in a second-hand store in Winnipeg, and asked what it was, but nobody knew. Most of the bystanders seemed to think it was for ferns. Among the other items in the room was a long sarcophagus with the lid off, lined with battered tin, and enclosed in a dark brown wooden case. It was very like an Egyptian coffin only older. It had a small hole in the bottom at one end, and a tap hung over it. From certain angles it bore a slight resemblance to a bathtub. I had no idea what it was for.

Above the sarcophagus was a terrifying contraption of metal. A large round cylinder, with a great entanglement of pipes, tubes, taps, valves, gauges, retorts, and gears coming out in all directions.

I turned on the tap over the sarcophagus, and to my surprise, water came out (cold of course). I cupped my hands and managed to scoop several handfuls over into the basin.

I washed a layer or two of dirt off my hands, and started in on my face. With my eyes closed I misjudged the distance to the basin and hit the edge with my hand. It tipped down and the water poured all over my trousers. Apparently it was loose in its moorings, so I investigated and found it could be picked up. This was encouraging, so I filled it under the tap and had another wash.

When the time came to empty it I didn't know what to do with the dirty water. Some of the earlier spill had missed me and run down onto the floor, and I was strongly tempted to pour the rest there too, but finally emptied it down the hole in the sarcophagus. It all disappeared, to my relief. I congratulated myself on a successful, though trying wash.

A much more trying experience took place later when I found out more about the gadget over the coffin.

I decided I had to take the bull by the horns and do something about the hot-water situation. After all, there comes a time when a towel won't take off any more dirt. So I got up bright and early and went downstairs and hollered for Miss Hards. She emerged from her sitting room in the cellar with a cup of tea, and after the usual formalities about the weather got down to business.

"About hot water," I started.

"Oh yes. You get it from the geezer, you know."

"Oh. I see." I was a little surprised at hearing slightly out-of-date North American slang in Greenwich, and was also a bit perplexed.

"Where is he?"

"Who?"

"This gentleman you said I get the water from."

"I didn't say you got it from a gentleman. I said the geezer."

This stopped me cold.

"The what?"

"Geezer. G E Y S E R."

"Oh, guyzer! I didn't know you had one of those in this country. Over in the park I suppose. That'll be fun! Imagine hot water shooting right up out of the earth. There's one in Yellowstone Park in the States too. Called 'Old Faithful.' Spouts off at regular intervals. What time does this one work?"

Miss Hards edged towards the door. There was a sticky pause.

"No," she eventually got out, "upstairs. In the bathroom. I'll show you how to light it."

Up we went, she keeping a safe distance from me, I completely baffled.

"That thing," she said, pointing to the tank with all the gadgets sticking out. She spent the next half hour or so showing me how to work it.

First you turn on a certain tap. This starts a violent gurgling sound but nothing happens – it is very important to turn this tap on first because if you don't you'll blow the house up. A full ten seconds later a stream of water starts flowing out of a narrow curved tube and falls down into the sarcophagus, the spray sprinkling everything at that end of the room.

Keeping as dry as you can, you light a match, and turn a smaller tap. You hold the match in front of this tap and if there is sufficient pressure in the gas mains a thin blue flame appears. If there isn't, the whole district blows up. This time we were lucky. Then you pull a lever which turns on the main gas supply and also swings the flame in under the tank. There should be a loud rippling whup and a whole lot of little blue flames start playing on a coil of pipe. If nothing happens, the pilot light has gone out and you either turn the whole thing off or else get asphyxiated. Under Miss Hard's expert touch the thing worked.

Then you turn on the ordinary tap that flows into the coffin for comparison purposes. (Miss Hards called it a bathtub.) In a few minutes you can distinctly feel a difference in the temperatures of the water from the two taps. This means that the water is now hot, so you fill up the basin.

Then you have to turn the geyser off. First the lever, then the pilot tap, then the main tap, in that order. Ten seconds after that the water stops splashing into the "bath."

If you turn everything off in the right order, all is well. If not, you are either blown up or gassed.

This process, together with the associated risk, is considered by the English to be a necessary part of getting washed.

It is the experience gained in childhood at geyser-lighting that has given England the world's most daring bomb-disposal experts.

I never lit a geyser by myself; I did all my hot-water washing at the college. I felt that London had suffered enough.

CHAPTER SIX

Ruins Ancient and Modern

On Saturday I took the train into London to continue my explorations, complete with huge map and aged guidebook. Old damage was very visible along the route. It passed close to several of the great dock areas that had been the scene of the daylight raid on 7 September 1940, when all the district had seemed ablaze. There were other smaller areas where house after house had been burnt out or blasted to pieces. One church, built at about the time of Sir Christopher Wren, always amused me in a macabre way. Instead of a traditional tower or steeple, the architect had seen fit to erect a tall, Grecian column with an Ionic capital. It pointed to the sky, holding up nothing but air and one or two birds' nests. The church was burnt out and empty, with just the four walls and this column remaining. I always used to chuckle at this amusing sight. It made me feel a little better about some of the eccentricities of our time when I realized that folly was not the preserve of the twentieth century. It stood there for years after the war, but to my sorrow was eventually torn down.

Daytime traffic in London was heavy with double-decker busses, taxis, and trucks, or lorries. Passenger cars were scarce due to severe petrol rationing.

The city was teaming with uniforms of all Commonwealth services and those of our allies – Americans galore, together with those gallant men and women who had escaped from the Continent. Walking along any sidewalk in the West End, the theatre district, one jostled against majors of the Grenadier Guards, Canadian army nurses, wing commanders of the Royal Australian Air Force, captains of the Fighting French, stokers wearing HMS on their sailor caps (ships were never named), sergeant pilots from Bomber Command, Wrens, WAAFS, WAACS, women in khaki from artillery batteries, rear admirals with rows of ribbons, aircrafts-men second class from the Royal New Zealand Air Force, American GIS, Waves, and members of the British Women's Land Army. These wonderful women passed up the comfort of home life and the glamour of the armed forces to serve as volunteers doing farm work. The hours were long and the work back-break-ing, but each woman relieved a man for fighting.

There were few young men to be seen in plain clothes, most male civilians being well over middle age. Ancient London taxi cabs could often be seen jammed full of American GIS cruising around town, waving to everybody and seeing the sights, which, judging by the number of wolf whistles emanating from the careening arks, included every woman they passed. These GIS would almost always be white. From time to time one would see a cluster of black American soldiers standing on a street corner chatting with small boys of the neighbourhood. The children were fascinated with their dark faces, strange accents, friendly manner, and chewing-gum handouts.

Our course started on Monday, 22 May 1944. The principal teacher was Instructor-Commander S.W.C. Pack, RN. About forty and handsome, he epitomized the adjective "pusser," naval jargon for supercorrectness, "to the manner born."

The spring morning was bright but chilly, with the feel of Iceland in the air. As time passed I became colder and colder until I could hardly write. Just as my hands were turning blue, Pack said in a loud clear voice in upper English, "It's getting a little stuffy in heah. Will someone please open the windows?"

I laughed at this witticism and thought the course would be fun – Pack had a great sense of humour. Then, to my utter astonish-ment, a couple of my English classmates promptly went over and opened the windows. We grabbed our papers to keep them from blowing off our desks in the icy draft. During the lunch break I ran the mile round trip to my room and put on my heavy-duty sea-going long underwear, which I wore for the next three days and nights. This was my first, but unfortunately not my last, encounter with British nuttiness about opening windows.

Pack had served as met. officer in the aircraft carrier *Formidable* during the early years of the war and had been in the Battle of

Matapan. He had released a pilot balloon at night to record upper winds and cloud base. A gust caused it to catch on the top of a tall crane on the flight deck. Its light was an open invitation to any Italian submarine in the vicinity to fire a torpedo at the giant ship. Our instructor-to-be climbed up the crane in a stiff wind and doused the light before trouble could develop. Pack's book *The Battle of Matapan* paints a vivid picture of the action.

I was the only trained meteorologist in the class, so much of it was simply review for me. I had no outside studies, and our weekends were free. In the weeks that followed I spent as much time as I could exploring London, which fascinated me. I had childhood memories of arriving there in 1926, the evening before the infamous General Strike began. We had been in one or two riots but had seen many of the famous sights despite the strike.

My fellow Canadians in the course were not overly interested in architectural sojourns and the English either knew London too well or didn't care, so my exploration trips were usually solos. But I never travelled in solitude. I found the English extremely friendly, contrary to popular repute, and most helpful whenever I asked directions. In fact, they were so courteous and eager to help that they frequently rerouted themselves and escorted me to my destination. Restaurants were invariably crowded, and a single person was usually asked to share a table with one or more strangers. In the wartime atmosphere they were never strangers by the end of the meal.

On one trip I found the Guildhall. This medieval banqueting hall had been used for centuries by the lord mayors of London for ceremonies honouring distinguished heroes. In 1940 I had seen newsreels of Churchill, then First Lord of the Admiralty, making a rousing speech to the officers and men of HMS *Exeter* after their defeat of the German pocket battleship *Graf Spee*. The hall was bombed a short time later, the roof collapsed, and the windows were blown out. But here it was, the walls still standing, with a temporary roof giving shelter from the elements. The door was open, so I entered the deserted building. The windows were boarded over, but a few shafts of sunlight came through the chinks and lessened the gloom. When my eyes grew used to the dark, I could make out long tables ready to host the next guests, whoever they might be. At the end of the room was the raised dais where I had seen Churchill in the newsreel. It was an extraordinary feeling being there in this famous place, all alone with the damage and history.

Westminster Abbey had also been hit by two or three bombs but one could walk around the famous nave as well as the damaged cloisters, which the public rarely visits. In WWI a bomb had landed in the Dean's Yard without exploding, but the noble building had not been so lucky in this war.

Sir Christopher Wren had designed a number of churches in the centre of London after the first Great Fire in 1666. Many of these had been burnt out by incendiary bombs; walls and towers still stood, but the insides were scarred and empty, usually with some weeds or pretty flowers contrasting with the destruction all around. Some of these ruins are still to be seen, fifty years later.

I was shocked to discover the area known as the Temple, where England's most distinguished law firms cluster around the Church of the Knights Templar. Built in 1185, this round Norman building with its heavily carved doorway was the oldest I had yet seen in Britain.

The walls and doorway were intact, but the interior was destroyed. At the time I wrote as follows:

Beyond I could see more evidence of German bombs – several fine buildings completely destroyed. The rubble was cleared up and wooden fences kept passers-by away from dangerous walls.

Oliver Goldsmith lay undisturbed beside the church close by the unidentified Knights Templar who returned from the Crusades ...

The Middle Temple Hall was badly damaged. It was built in 1572, and Queen Elizabeth [the First] attended a performance of [Shakespeare's] *Twelfth Night* in it.

A nearby small house containing lawyers' offices was wiped out; nothing remained but the front door and the name plate.

Johnson, Boswell, Charles Lamb had all lived near by and often must have strolled along the same walks that now lead from ruin to ruin.

This was the first time I had seen irreparable bomb damage. Liverpool's business streets can be rebuilt in a more modern and up-to-date style, but nothing can ever replace the Temple. The Germans have lost just as much as we have as a result of their destruction of this shrine of chivalry and justice. Parts can be repaired, others can be imitated, but the world has lost forever something that it needs. I walked away feeling a deep sorrow and also a disgust at this vandalism.

Years later I visited the area with Janie, my wife. Much of it had been rebuilt with incredible skill so that it was hard to remember the scene described above. Our great friend Peter Bevan, who had been my senior met. officer at Yeovilton Naval Air Station and was now reestablished in his law offices, showed us around. We had the honour of being taken to lunch in the exclusive Middle Temple Hall, now miraculously restored to its former glory. The decor, although obviously not the original, nevertheless gave the feeling of freshness and glitter that must have greeted Queen Elizabeth at that celebrated première.

One day I walked a block or two north-east of St Paul's and came upon a staggering sight: just as in Liverpool, a huge area of emptiness – basements only, with street after street in between, and one or two burnt-out buildings. In the distance could be seen the shell of St Giles Cripplegate Church, where Milton lay buried and Morley had been the organist in the time of the first Queen Elizabeth.

This area was much larger than the one in Liverpool and had been destroyed mostly in that one terrible raid on the night of 29–30 December 1940. It was the old City district, once full of financial offices and other business premises, with practically no permanent residents. The story of the fire, which is often called the second Great Fire of London, has been told many times. But I have not seen an account of the emotional effect of this silent area in the years that followed. There were no craters, no rubble, and no evidence of recent damage, just block after block of open basements. Here and there gardens had been started, and on one a sign proclaimed victory in a bombsite-beautification contest.

Several times in the weeks that followed, I walked through this deserted area in the quiet, sunlit evenings and felt that I was walking on a great battlefield, as truly I was.

CHAPTER SEVEN

Where's the Pilot?

When I arrived in London I had no idea what to do if an alert sounded. We had had a few practice blackouts in Canada, when everyone was required to go indoors and hide. I decided that I should find out how to behave in a major-league air raid. I explained my dilemma to a policeman.

"Just carry on, sir," he replied.

"You mean we don't have to go to a shelter?" I asked.

"Good 'eavens, no, sir. There's no telling what you might catch in one of them plyces. If we took shelter every time the sirens went nobody would ever get any work done," was the emphatic reply, followed by a more casual afterthought. "Of course, if things get close you might pop under cover somewhere."

I went on my way wondering how one knew when things were going to get close. In due course I found out.

By Monday, 12 June we had almost forgotten about Hitler's V-threats. The armies, whose progress we followed admiringly, were making good headway in Normandy except around Caen, and there had as yet been no unpleasant surprises.

After a normal day at the college I returned to my room and went to bed. Some weeks afterwards I wrote as follows:

A week passed with no sign of enemy air activity, so it really did look as if Hitler were concentrating all his aircraft on the Battle of Normandy. I thought that the most we would be likely to get over London would be a stray reconnaissance plane or a few nuisance raiders. People told me that London often had alerts when aircraft came inland, even when nothing was heard in the city itself.

At a quarter past four on Tuesday morning, 13 June, one week after D-Day, I was violently awakened by a very loud shrieking wail. I sat bolt upright in the dark, and in a second or two realized that it was an air-raid siren. The sound rose and fell several times in a hoarse glissando of parallel minor thirds. Beyond, others were rising and falling at different times. It was the weirdest, most terrifying yowling imaginable. After only a few seconds through the noise of the sirens I could distinctly make out the distant pump-pump-pump of gunfire. I hopped out of bed and went over to the window. Pushing back the curtain I could see the beams of several searchlights exploring the sky to the south. The sirens' screaming died down to a throaty gargle, and the gunfire increased in volume and intensity but was still not loud.

I was quite scared. The enemy had reached the outskirts so suddenly that possibly something exceptionally heavy was in store for us.

After about two minutes the firing ceased and was replaced by a very ominous silence, broken only by the absurd toot of a shunting engine and the shrill voice of a woman air-raid warden yelling, "Put that light out!" As I didn't hear anyone moving about in the house I decided that everyone else was staying in bed.

I regret to report that my knees shook and I felt limp. Being awakened at four in the morning by anything sudden is startling, and the noise of the sirens is bad at any time. All my life I had been afraid of air attack, and here it was at last.

The darkness outside was just beginning to turn grey, but still the back gardens were darker than I had previously seen them due to the long evenings. I could just distinguish trees and houses, walks and grass.

Everything was silent now except my teeth. I remembered what the policeman had said about carrying on, so as things seemed to have stopped happening I went back to bed. I steadied down quite a bit, by which I mean I stopped shaking the attic.

Just then all the devils in hell started their vocalizing, as the sirens sounded the All Clear. They scoop up to a high loud note, hold it for some time, and then slither down the other side. The All Clear lacks the character of the alert, but is harder on the ears. I found it very good for the nerves, though, and in a few minutes felt quite courageous. I could probably have swallowed if I had had any saliva.

For fifteen minutes I thought about My Experiences of Air Attack, and decided that I'd put on a pretty poor show. I was quite ashamed of getting in such a dither. My heart disentangled itself from my teeth and slid back into place, still keeping up a brisk *allegro vivace*.

I decided that it would be pleasanter to be killed than to get scared that much. Whereupon the sirens went again and once more my heart closed up smartly to action stations behind my front teeth, shifting into *prestissimo*.

However, I was now an old hand at air raids, and as I heard no firing I stayed in bed. The last growl of the nearest siren died out, and the silence became very intense. Not a sound this time.

After several minutes I heard the engine of an aircraft, which within a few seconds became very loud. It had a harsh, snarly tone very similar to a Harvard training plane. He was obviously alone and quite low, and I thought he must have been one of ours as I could hear no firing.

Suddenly a terrific clatter broke out as a very near AA [anti-aircraft] battery opened up. I nipped over to the window but the aircraft had passed to the east of us, heading north. The engine cut, and I thought, "They've hit him." The firing went on and in a few seconds a rather distant, louder explosion told that the aircraft had crashed. The firing stopped, and the silence was once more turned on. The whole incident had taken about half a minute.

This had been quite exhilarating, and the idea of all London shooting at one stray German appealed to me, particularly after he had passed Greenwich. I went back to bed feeling that was that, and now I had something to tell my grandchildren about. In a few minutes the All Clear went, but I was braced for it and pulled through better …

The next day the papers reported that there had been slight enemy air activity over London and that one aircraft had been shot down, crashing at Mile End with a full load of bombs. It had demolished a row of houses and caused a diversion of traffic on a line of the LNER [London and Northeastern Railroad]. One paper had stressed the unusual blast of the explosion, which had broken a shop window half a mile away.

At breakfast we all talked at once and told our own stories of the affair. Several said that the aircraft had passed almost directly over the college, and that the engine had stopped when over the river.

I was duty officer of the class that day, which necessitated staying in the classroom during the normal lunch hour. A charwoman was sweeping up, and we chatted about the raid. She told me one detail I hadn't heard before – she said she had seen the aircraft and that its lights were on when it flew over. This was the first suspicion I had that the incident was not as normal as it seemed.

When the rest of the class came back after lunch some of them told me additional bits of information they had picked up. Several people backed the statement about navigation lights and said the pilot probably didn't know they were on. Others said the aircraft was on fire, and some that it was carrying one white light. Everyone agreed it was flying low, very fast, and strangely enough took no evasive action whatever – it flew straight and level until it was hit. We decided the pilot had probably been killed, wounded, or dazed, or might even have bailed out beforehand.

However, that night was quiet, and by morning the incident was closed.

Next day the incident opened up again. A buzz went around that the searchers still hadn't found the pilot's body in the wreckage. This strengthened the belief that he had jumped when the aircraft caught fire and had left it on the automatic pilot.

The next day [Thursday], another buzz went around, this one behind backs of hands – there never had been any pilot. The aircraft was a robot similar to the controlled pilotless glider bombs that the enemy often employed against shipping. This was more or less a guess and was not given much of a hearing in view of the complete lack of evidence, but it provided food for thought.

CHAPTER EIGHT

The Doodlebugs Attack in Earnest

Thursday, 15 June on first acquaintance appeared to be just another day, but before it was over, things had once more started happening to London in a big way. After dinner that evening I played some tennis. We had to clear anti-aircraft shell fragments off the court before we could begin, jagged reminders of the nocturnal visit two nights earlier. I went home and wrote a letter. According to my 1944 account, the evening was cloudy and not quite dark.

I went to bed soon after eleven ... About two minutes later, as I lay letting my mind gradually run down, I thought I heard a very distant air-raid siren. Just then the last tram went past and drowned out the sound, and I decided that I had either imagined it or else had heard a truck doing an imitation.

At about twenty past eleven our neighbourhood siren started up and was soon joined by all its relatives. I had apparently heard the warning from another part of the city.

I got up and went to the window, excited, tense, a bit scared, and slightly curious. In a minute or two, distant firing began, and a few searchlights aimed over to the east, beyond where I could see. The firing grew louder, and I became aware of that same hoarse sound of an aircraft engine, first very faint, then becoming louder. The searchlight beams slowly swung back until I could see them converging in one brilliant bluish white glare, into which crimson balls of tracer floated from the ground. As it came nearer I could make out a small yellow light in the centre, the colour of a match flame. It was approaching faster than most aeroplanes and rapidly moved across the sky, a mile or so to the south, keeping on a straight and steady course. The firing and the engine noise both became quite loud, and the red tracer balls streaked at the aircraft. When it was about three-quarters of the way across the sky the light went out. I thought he'd been hit. In about five seconds the engine sound stopped. In a few more there was a heavy explosion, and a tall column of black smoke boiled upwards, dark and menacing against the lighter sky.

The firing subsided and the searchlights were switched off. I felt quite steady and was thinking very hard. The straight course of the plane – the light – no body in the wreckage of the other one – it looked very much as if this certainly *were* a secret weapon.

Distant firing began again in the east. I hastily dug out my binoculars. The searchlights came on, aiming to my left.

Another blue white patch of brilliance moved in sight, again with a central dot of yellow and rising globules of crimson. I focused my glasses and followed it across the same path as the other. The light appeared to be long and narrow, with a ragged tail. This time it passed out of sight to the right. The engine cut, and in about ten seconds a heavy but distant explosion rattled the windows, followed by silence.

I was now convinced that these were pilotless aircraft, but I thought they were an adaptation of glider bombs, which are controlled and aimed by radio from a nearby aircraft. I thought the light was to show the guiding plane where the other was and that its elongated appearance and ragged end were an optical illusion due to the speed. I presumed the parent plane was hiding in or on top of the cloud layer. This of course proved to be wrong. [Later evidence indicated that some of the early bombs were, in fact, followed by standard aircraft to track their course. In the autumn many were fired from aircraft that carried them underneath until released.] The logical thing seemed to be to shoot the aircraft down before it reached its target and did severe damage.

There was no sound in the house. Outside I could see a few of our neighbours making their way in the twilight to the Andersons [shelters]. At that point there were footsteps outside my door, and someone knocked. I said, "Come in," and Miss Hards entered. It was dark in the room, but I could see she was wearing a dressing gown and her hair was secured in a net cap. Her small face was white and her thin lips tightly shut.

"Mr Mackelerheran, would you like to come down to the basement with us? We're all down there and I really think you should be safer there than up here." Her voice trembled ever so slightly.

"Thanks so much, that's very nice of you to come up and ask me." I noticed mine was shaky too and my mouth was dry.

The firing started again, and another aircraft appeared in the east and moved across the same path, the anti-aircraft crashing at it. We looked out together. There was a touch of the uncanny in the way the light glided past. "Did you see the other two?" I asked.

"I saw the first one. I say, what do you think of these things? Rather queer, aren't they?"

"I'm sure they're pilotless. Probably controlled by radio," I answered.

The light vanished, and we stepped back from the window. Another big bang a mile or two away.

"New Cross again. Or maybe Lewisham," said the landlady. "The first one was also on New Cross."

"Is that what that district's called?"

"Yes. Next borough over."

She said she'd never seen enemy bombers flying alone on a low, straight course before, and never with a light. I told her what I had read of glider bombs and explained my theory. It impressed her. That made two of us.

I noticed that after a few sentences both of our voices became quite normal.

Things were quiet outside again.

Miss Hards was not mechanically minded. She was the product of an age that didn't give a very good scientific training to most men, and little or none to women. Yet here she was discussing pilotless aircraft and secret weapons in the same everyday manner she would use at the grocer's. She wasn't flippant or blasé, or even casual, and she certainly wasn't attempting to be humorous. I could detect in her rather rapid gestures and the tightness of her face muscles an underlying uneasiness. But on the surface she was serious and interested.

I don't know if we realized then that we were witnessing the beginning of a great episode of history. I think we were more concerned with details, like where the parent aircraft was and whether we should go down to the cellar.

I was glad of her company and was rather touched by her coming all the way up to the attic during the raid to see if I was all right. I was quite excited at what we had seen, and wanted to talk it over with someone. A thousand ideas kept hurtling through my brain. Could we jam the wavelength? Surely our fighters'll be able to get them. I wonder how many he can fire at a time? Our AA fellows certainly did well. Where's my helmet?

It was now about 11:40 and it looked as if it were all over, for the time being at any rate. I told Miss Hards that I thought I'd go back to bed

unless things got worse. She made me promise to go down if I wanted
to and convinced me that I'd be quite welcome. Then she left and I went
back to bed, expecting the All Clear at any minute.

At about 12:00 the firing recommenced in the distance and died away
without getting any nearer. A little while later it started again, and in a
minute something very close opened up with an astonishing clatter. The
aircraft was much louder this time, but I couldn't see it. After a few
seconds hundreds of little pieces of flak [fragments of anti-aircraft shells]
came down around us. It was a very pretty sound filling the air with a
fairylike tinkling, broken occasionally by the thud of a piece on our roof
or a metallic ring as a larger chunk hit the pavement outside. It lasted
several seconds and was almost celestial in its beauty. Heavy flak makes
more noise, of course [but this was from the light guns as the aircraft
were too low and fast for the heavies to operate].

Again there was a knock at my door. This time it was John, the Irish
lad from the next room whom I had never met. It was a good deal darker
now. I could just see that he was of moderate height but very thick-set
and dressed in a suit and sweater.

We introduced ourselves and started to work on the secret weapon.

"Did you see that last one?" I asked.

"That I did. Went over Greenwich heading north."

"Did we get him?"

"Yes, I think we did that. He seemed to crash Poplar way."

"Some fun, eh?"

"Yes. I wonder how long they'll last."

"Well, the other day the raids packed up shortly after four, when it
was starting to get light. I don't suppose they'll keep coming much past
that," I replied with my usual clairvoyance.

We discussed the business and agreed that the whole thing was very
interesting. But we felt that the rumours about them that had issued from
Germany were exaggerated, and we were surprised that Hitler would
use a secret weapon that could be seen so far at night. Our tendency was
to underestimate their value. We didn't realize that their speed was so
great and the idea of jet propulsion didn't occur to either of us. [It also
never entered our minds that the Germans would be so wasteful of
material or so cruel that they would invent a weapon that could not
possibly be aimed at a target smaller than a huge metropolis.]

There was some more distant firing, and we saw searchlights hunting
far to the south, but nothing came near. We talked on and on about
Ireland and Canada, the navy and the Blitz. He had lived for several

years in the same room and spoke very highly of Miss Hards and Billie. He told me a lot about the Blitz and also about the spring raids that year, one of which had been unpleasantly close.

A pilotless aircraft raced across the sky to the south with half of London flying up at it, but it kept on going out of sight. Most of the flak went astern. It was later announced that the gunners had found it very hard to keep their guns trained on such a fast-moving target. Actually, most of the aircraft were crashing of their own accord, but at the time we thought they were being shot down. It wasn't until the anti-aircraft people began putting up a fixed barrage that their fire became effective. John went downstairs to have a chat with the ladies.

Later another came near us and the noisy battery rattled the window once more and sprinkled some flak around the neighbourhood.

I tried to get to sleep but each time I was almost under, the firing started again. It never lasted long and wasn't terribly loud, but on the other hand it wasn't soothing either. I kept my helmet handy and put it on once or twice when near ones came over.

John came in a few times more and I went in to his room a couple of times to watch aircraft flying north of us. He said that the ladies were all right, and that George II had moved down to a couch in the ground-floor dining room.

Time dragged by in the way it does when you can't get to sleep. There were long periods of quiet, some as long as half an hour, but I never could doze off. When the luminous dial of my travelling alarm clock said 3:00, I thought, "Dawn in another hour. They'll surely stop then."

The hour crawled by. At four it began to grow perceptibly lighter. At half past it was light enough outside to see objects a block away. At five it was broad daylight. But the aircraft kept coming over, sometimes isolated from each other, sometimes in succession.

From time to time a nearby gun would fire a few rounds off by itself, without any apparent target. This was hard to explain, but I supposed they had a radar contact in the cloud or maybe they just liked the noise.

I was feeling very groggy by this time and had lost all fear or interest. The tennis of the previous evening had tired me anyway. It seemed years ago, when I thought back to it. I thought about my wife and mother and wondered how long it would be before they heard about the excitement. The letter I had written still lay in its envelope on my bureau.

I felt I should be awake when aircraft came near so that I could get under the bed and put on my helmet in case the roof came in. Finally, I got fed up and decided to go downstairs and get some sleep without worrying about the roof.

I climbed into some old gray flannels and a sweater and found my way down the three flights of stairs. The cellar consisted of a bedroom at the front, a small kitchen under the stairs, and a moderate-sized sitting room at the back. It had one average-sized window, the top of which was above ground level, the remaining two-thirds opening into a narrow trench or well a foot or so wide, walled up with bricks. It admitted light and air and, as we later found out, blast.

In the room were two small couches, an easy chair, a dining table, and a collection of plain chairs, footstools, china cabinets, bookcases, etc., all of which made excellent tank-traps in the dark. On the walls and mantle was an assortment of Victorian relics.

Miss Hards and Billie were down there, lying on the two couches. Both were awake. They immediately got up and started moving things around to make me a place to sleep. Cushions and rugs appeared in large numbers and I settled on the easy chair with my feet up on another. They tucked me in like a baby and then went back to their couches.

We talked for a while about the secret weapons, but we were bored with them by that time and soon stopped. The secret weapons, however, didn't. They kept coming at widely spaced intervals, and I began to wonder if they'd go on all day.

I managed to sleep for about half an hour but woke up as the AA opened fire again. At about seven we all gave it up as a bad job. I went upstairs and started to get dressed, feeling very dim.

When I was getting dressed I heard an aircraft coming over and stuck my head out of the window to have a look at it. It was flying in the cloud, but as it went overhead I caught a glimpse of the trim grey aircraft and long thin body that was to become so familiar to the people of southeast England. Its wings were rocking drunkenly as it scooted over. It had probably been hit and its gyro was affected. I heard it come down a mile or so away.

I had a wash and a shave and broke out a clean collar to keep up my morale ...

Just then Miss Hards brought up a cup of tea. She didn't see how anyone could start a day after such a night without a cup of tea.

All this time there was no sound of raiders, and the morning traffic noises were beginning as usual. The rooster next door sounded a little hoarse but was nevertheless on the job.

When I went downstairs to go out to breakfast Miss Hards came up from the kitchen to see me off. She was now neatly dressed, but her face showed signs of strain and lack of sleep. She wanted to make sure I took my helmet in case of falling schrapnel.

We argued about this for a moment. "Grandfather" said five past eight, which meant it was a quarter to. Distant firing started up again.

"Oh dear, there's another," said Miss Hards with a sigh. "You'd better step inside for a minute."

We went into the front parlour, which I had never been in before, and the anti-aircraft became very loud, almost drowning out the snarl of the aircraft.

"I'd better draw these shutters," said the landlady, taking one step towards the window. At that instant there was a very loud explosion, shaking the whole house. We distinctly saw the large windowpane bulge in and out very rapidly about a quarter of an inch without breaking, just as if it were rubber. A tiny stab of pain jabbed both my ears.

"That's a close one," she said, as the flak tinkled down.

We went outside the front door. A thick pall of brown smoke was drifting along the street from the western end.

"It's fallen at the bottom of the road," she said ...

People were coming out of all the houses and looking down the street. I nipped upstairs and put on my helmet and then started along to see if I could be of any help.

Two houses away from us the front window had broken and fallen out into the little garden. Several other windows along farther were broken, and some pieces of glass had flown as far as the sidewalk, crunching under foot as I walked. [As in many cases when a bomb exploded at a moderate distance, glass was broken by the low pressure or suction wave that immediately followed the first blast. It would fly towards the bomb. Our window had shown this effect but was far enough away to survive the double thrust.]

Most of the inhabitants of Ashburnham Place were standing outside their houses, some on the steps or on the lawn, others on the sidewalk or in the middle of the road. There were a few very old ladies huddled in shawls, dazed and bewildered. Plump housewives in dressing gowns, dresses, and aprons, or even slacks, shouted across to each other.

"Wonder where 'e 'it?"

"Wasn't it a terrible night – never slept a ruddy wink."

Smaller children sucked their thumbs thoughtfully, while older ones argued about how fast 'e was going.

"Coo, all of five 'undred, betcha."

"Naw, yer barmy."

A middle-aged man in shirtsleeves gobbled a piece of toast, afraid of being late for work. A few other men were walking ahead of me towards where the bomb had landed. The smoke stopped blowing across the street and cleared up.

There was a funny atmosphere in Ashburnham Place that morning. Curiosity, awe, headaches, worry, perhaps a bit of fear, and a peculiar mingling of reverence and suspicion of something the locals didn't understand. The little street had seen some bad nights in two wars, but this one was a bit rum. Ashburnham Place wasn't afraid exactly, but it had the creeps. It preferred aircraft with pilots to these bloody queer things.

I turned on to Greenwich High Road. Traffic was moving normally and a number of pedestrians were about. The bomb had landed a good deal farther away than we had thought; there was no new visible damage except a number of broken windows.

I came to the Miller General Hospital, a new brick building in modern style. Most of its windows were out, and the sidewalk was covered with broken glass. As I passed the entrance a woman got out of a car holding a bloodstained handkerchief to her eye. She was followed by a poorly dressed man carrying a little boy whose face was streaming with blood. They both went into the hospital and the car drove away. It was a taxi, and its "for hire" sign was up, showing the meter hadn't been turned on.

As I walked on I came into the shopping district. All the large front windows were out, lying in millions of pieces on the sidewalk and road. In places the broken glass was ankle deep. In front of almost every shop a woman was sweeping up. It was an unbelievable sight, all those people casually sweeping the glass into neat piles in the gutter while the raid continued. The sun was breaking through the clouds as the sound of tinkling glass filled the air.

I entered a sidestreet where the bomb had fallen, just as a truck full of firemen drove up. They piled out and trotted ahead of me, carrying axes for rescue work. [There was no fire.] Several very old women were sitting beside the street on kitchen chairs provided by neighbours. They looked bewildered and unhappy. One or two were bleeding, and a couple of nuns were bandaging and comforting them.

A number of civilian men and women in helmets were scattered along the road ahead of me, mostly standing around wondering what to do. There was a great deal of dust in the air, shining yellow in the sunshine.

Off the road was a narrow driveway leading into a court where the aircraft had fallen. A wooden fence at the entrance was blown outwards. Glass was everywhere.

The courtyard had apparently once contained several slum houses, but all that remained were piles of rubble several feet deep – crumbled plaster, broken bricks, and roof slates. White dust lay thick over everything and filled the air so that it was hard to breathe. Many firemen were digging in the rubble searching for people.

One woman was directing the firemen and giving particulars of the residents to a policeman and an air-raid warden. She was about thirty-five, of moderate build, and was wearing a dark red dress now whitened with dust. Her face was completely black with soot from the explosion ten minutes before. When she talked her teeth gleamed white. A trickle of sticky blood oozed from a cut over one eyebrow. The blood, too, was darkened by the soot. Her manner was calm, capable, and friendly. As she pointed to one wrecked house I heard her say, "Then there's old Mrs Jackson in number 16B – that's that one," as if she were directing a new postman.

I'll never forget that woman. To me she symbolized everything that I had read and heard about England. She wasn't an exceptional heroine performing a spectacular rescue feat; she was just another bombed Londoner helping the authorities do their job.

I went over to where most of the firemen were searching. The rubble was three feet deep and nothing was left standing but one small corner of the house. Two firemen were looking around the wall at something that I couldn't see. A third stepped over towards me and shouted, "For God's sake, somebody get an ambulance. There's a woman in 'ere with 'er guts 'anging aht." A civilian doubled out to the street to get one, while the firemen picked up a door that was lying around to use as a temporary stretcher.

I reported to the fire chief and asked if I could be of any help, but he said that there seemed to be quite enough people around. I started back through the lane just as the ambulance drove in. There were several people standing in its path, one of them being a very old man bent almost double and leaning on a stick. Someone took him by the arm and shouted in his ear, "Look out, Mr Giles. They want to get an ambulance in here." The old man irritably shook off the other's hand and snapped, "Let me alone. I'm old enough to look after meself." He then hobbled into a doorway and the ambulance passed by.

I went out to the street and met an elderly lieutenant colonel of the Home Guard, just arriving. He was in battle dress and a helmet, and three rows of medals showed he had been in both the Boer War and World War I. He was very scrawny and stooped. On one hip was a revolver in a holster. He was the senior officer present, so I saluted him and reported the situation.

It was almost time for work, so as there seemed little point in my remaining any longer I caught a tram back home.

CHAPTER NINE

The Battle Continues

When I reached the college I was relieved to find that the class had been postponed for one hour and I was in time for breakfast. Needless to say, we all had various accounts to relate and did so with vigour. Those of our course members who lived in western London reported seeing no effects from the night's attack, the only visible damage being the scars left years earlier by the Blitz.

At about 9:30 the All Clear sounded, only to be replaced by a succession of alerts, aircraft noises, distant explosions, and All Clears, not always in that order. An examination had been scheduled for that morning and of course was held, this being the Royal Navy.

That afternoon the announcement was made to the world at large that southern England had been attacked by a number of pilotless aircraft and that some damage had been caused. The term southern England was used throughout the months ahead to avoid giving more specific information to the enemy.

After the war we learned some of the official statistics. The first V-1 arrived at 4:18 A.M. on Tuesday, 13 June. It crashed at Swanscombe, west of Gravesend, about fifteen miles east of Greenwich. Six minutes later another struck at Cuckfield, Sussex, well to the south of us, causing no damage. The third, soon afterwards, hit the north-west side of the LNER railway bridge across Grove Road, Bow, near Mile End Underground station, killing six and injuring forty-two. This must have been the one that passed over Greenwich. A fourth landed at Crouch near Borough Green, Kent. I could never find which of these four caused the two alerts.

There is ambiguity in the postwar accounts. There were definitely two separate alerts at Greenwich a few minutes apart; during the first we heard distant firing but no aircraft, and during the second the pilotless aircraft was seen flying over Greenwich, doubtless on its way to Bow. The books say that it landed at Bethnal Green, but according to local people it crashed at Bow or Mile End, the next stop on the Underground.

Joseph Waters has supplied details concerning this bomb. His research and initiative led to the installation of a memorial plaque on the railway bridge over Grove Road marking the spot.

An interesting item that adds further confusion appeared in 1974 in the 16 June issue of the *Sunday Telegraph*. It read as follows:

THE VERY FIRST V-1? The article by Duff Hart-Davis last Sunday says that the first V-1 to explode on British soil was at Swanscombe, near Gravesend, on June 13, 1944. I expect this was confirmed from Official Records, but I suspect the Official Record is incomplete, for an understandable reason.

A doodlebug exploded in Hampshire long before June 13. I was on duty in the control room at Fareham on Whit Monday, [29 May] 1944. The hour before midnight was still warm after a hot day and there being no "Red" alert we enjoyed the fresh air outside the doorway.

The sound of a plane with an engine making an unusual noise was heard and then came into view and appeared to have its tail on fire. Moments after it passed over there was an explosion.

Wardens' post messages came in of a crashing plane, the crucial one coming from Park Gate, Sarisbury Green, that a plane had crashed on the tanks parked at the roadside and there were casualties. Services were dispatched and the incident reported to Region.

However, the Military (Canadian) took over and supplanted the Civil Defence. In the early hours of the next morning Region instructed me to delete from copies of all control room messages every reference to "plane" and substitute "bomb."

I argued on the evidence of my own eyes but was told that this was an order from Whitehall and was to be obeyed. Later I was convinced I was right and it dawned on me that the secrecy was to deny the Nazis of knowledge as to range and direction.

        — B.W. RANDS, Fareham, Hants.

In Kent County Council's *The County Administration in War* the following passage adds further ambiguity. The shells referred to were from long-range guns that had been firing at southeastern coastal towns intermittently since the fall of France.

The new phase of enemy fly bomb activity had opened on the night of Monday, 12th-13th June, 1944, continuing into the early hours of Tuesday morning. Maidstone and Folkstone were bombarded by cross-Channel guns. Thirty-four shells were fired at Folkstone, 26 of which fell on land, and widespread damage was caused to property. Maidstone, too, was shelled by enemy guns.

During this activity, what were reported to be a few aircraft operated in two phases, but only four incidents resulted, one in Bethnal Green where severe blast damage was caused, and other incidents at Swanscombe and Cuckfield, Sussex. The second phase began at about 5 o'clock on the morning of Tuesday, 13th June, the only incident ocurring at Crouch, near Borough Green. These turned out to be the first fly bombs.

This implies that the first attacks were on 12 June, some considerable time before the Crouch bomb, but the official records refer to 4:18 as the arrival time of the first V-1. At Greenwich the two alerts were within a few minutes of each other.

Postwar records indicate that in the twenty-four-hour period commencing on the evening of Thursday, 15 June, 151 flying bombs attacked England, launched from fifty-five sites in France. Seventy-three reached London. No wonder it had been a noisy night!

Although the theoretical aiming point for all V-1s was Tower Bridge in central London, they were very erratic and generally landed over a wide area some distance apart. The capital extended over 360 square miles, measuring about eighteen miles by twenty, so there was usually considerable space between "incidents." There were about eight million people in the London area at that time, the vast majority of whom had not yet seen their first V-1.

The cabinet met on Friday, 16 June and decided not to allow any men or equipment to be diverted from Normandy to defend London, which would just have to put up with it. This decision soon had to be modified. The cabinet in its infinite wisdom also decided that these weapons should be officially called "flying bombs."

We at first thought of the intruders as aircraft, then pilotless aircraft. After crashing and exploding, they were considered bombs. However, characteristic of British humour, they were dubbed "doodlebugs," a name that came from nowhere and was soon heard everywhere. It had some of the Londoners' bounciness and showed the amused disdain that the British felt for these much-ballyhooed gadgets.

One postwar explanation of dubious merit credited the term to New Zealand pilots, who named them in honour of an antipodean insect. They were also called "buzz bombs" because of the rasping sound of the engine. The German term V-1 was short for *Vergeltungswaffe*, or "Vengeance weapon." We didn't' call them V-1s until after the V-2s had started to sprinkle down among us. In the 1980s they would have been called cruise missiles. Indeed, watching the Scud attacks on Tel Aviv during the Gulf War in 1991 brought back vivid memories.

On the same day, Friday, the BBC announced that the friendly tones of Big Ben preceding the nine o'clock news would henceforth be prerecorded. They were afraid that the enemy would hear explosions or alerts and be able to judge the accuracy of their aim from such clues.

We soon learned that these creatures had a wing span of sixteen feet (later seventeen and a half), a length of twenty-two feet (later twenty-five), with a warhead holding almost one ton of high explosives. They were powered by a jet engine new to us called a ram jet or pulse jet, mounted in the tail. This emitted the yellow exhaust flames that could be seen at night. They flew over 360 MPH at a height of two to three thousand feet. They were launched from bases mostly in the area near Calais, in northern France, and also from many other hard-to-find locations along the French coast.

A spinner in the nose of the flying bomb calculated the distance flown, and when it thought it was over London a gadget in its navigation equipment was tripped, causing it to head downward. This in turn cut off the fuel supply, bringing on the deathly silence that we all came to know so well. Some simply ran out of fuel, with the same effect. Forty-four percent made their dive in silence, and fifty-five percent plunged part way with their engines roaring before the silence took over. Only four percent performed as they were designed to, power diving all the way to the ground.*

Sometimes the steering mechanism would misbehave, causing the flying bomb to make a U-turn or even a complete circle. This erratic conduct hatched a swarm of rumours about people who swore they had been personally chased by a doodlebug.

They were aimed completely at random and were entirely designed to terrorize the population. They failed to do so, but they got on our nerves.

Friday night was quiet compared with its predecessor. We never figured out why some days or nights were much worse than others. It certainly wasn't due to the accuracy of our gunners, who found it very difficult to train their guns on such fast and low-flying targets.

During the daytime on Saturday there were several alerts, when we sometimes could hear firing or explosions, followed by the All Clear. Other attacks were obviously well out of earshot and we just heard the sirens. Occasionally the big bang would come and then the belated alert would be sounded, the fast-moving aircraft having fooled the defence. Although most doodlebugs came singly, from time to time they arrived in clusters of three or four.

We frequently lost track of whether an alert was on or not, especially if we had been below street level. However, we soon realized that the general warning of aircraft over the coast was of little importance to us. We developed a survival technique of using our ears and taking shelter where we could when we heard a buzz bomb approaching. A cartoon appearing at the time showed a street full of people. In the distance is a tiny aircraft. Everybody is walking along normally, but each person has one huge ear aiming in the direction of the enemy.

That night I went to bed in my room, and once more the sirens started, this time later than usual. There was a loud explosion somewhere nearby, or at least so it seemed, but it was difficult to assess accurately distance with these miserable creatures because their lateral blast often made them sound closer than they were.

---

* These figures are taken from Richard Anthony Young's *The Flying Bomb*, 91. They add up to a hundred and three percent.

I decided that I should go out and help at whatever point the bomb had struck in case the rescue workers were short-handed. As I left the house I met my fellow boarder, John, the Irish longshoreman. We decided to team up. It was quite dark but I could tell that he was in working clothes. I was in uniform as required by regulations. We both felt that we should head towards where we thought the bomb had landed and we hurried in that general direction without any real idea of what we were doing. After a few minutes we came to a place where Greenwich High Road and New Cross Road converged, forming a little triangle with a low building on an island in the middle. Just as we reached there we heard the now all-too-familiar sound of an enemy air-craft droning its way towards us. Suddenly we saw its flame appear almost overhead. We flattened ourselves face down in the gutter beside this little house which seemed to offer moral support if nothing else. The engine, now very loud, stopped and we more or less held our breaths. After the usual ten seconds of silence, a mighty bang from across the river showed that the aircraft had landed well away from us, near the Isle of Dogs. We picked ourselves up and dusted ourselves off and realized that we could not get across the river to help. We wandered around awhile longer, but nothing else happened in our district so we returned home and went to bed for another restless night.

I was rather pleased with myself about this because my knees had not been rubbery, and I had felt fairly calm as I lay in the silent shelter of the little building. I thought I had grown up considerably since that terrifying first alert five nights earlier. However, some of my self-satisfaction was taken down a little the next morning when I walked past that intersection and found that the little old building was the entrance to a very odoriferous underground men's toilet. At least I was glad I hadn't been killed there.

CHAPTER TEN

Panic or Nonpanic?

Now Hitler made a great mistake. The Sunday papers and the BBC reported that his propaganda machine was screaming to the world that London was in ruins and that the population was pouring out of town in panic. This was considered quite hilarious by the fun-loving British public, most of whom at this point had not seen a V-1 or any new damage whatsoever. If Hitler and his propagandists had understood the English better, they would have said, "You haven't seen much yet. But we have a great supply of these devices and during the months ahead you will find them very hard on your nerves. Every few days you will notice a new familiar site blasted, every few days a friend of yours will be killed or made homeless. This will go on month after month." Instead, they screamed that London was devastated now and thereby gave the British a tremendous psychological boost.

After a couple of days the public was enormously relieved that this new assault did not seem nearly as bad as the earlier raids. Veterans of both the Blitz and the Little Blitz considered this child's play.

At about that time, a cartoon appeared in a London paper. In a suburban garden, a pleasant middle-aged woman is talking to her husband, who is seated in an easy chair. He is obviously irate. In the distance can be seen a tiny aircraft. She is saying, "You can't wring that pilot's neck, dear, because there *is* no pilot." This isn't the funniest joke that ever appeared in print, but it was at least topical and showed the revision in thinking that we were all going through as the world entered robot warfare.

On Sunday, 18 June, I went into town again, but despite the night-long alert there still was no new evidence of the attack to be seen from the railway. In fact, it was surprising how long it was before many Londoners saw their first damage.

I chose that morning to browse around the neighbourhood of the Tower of London, at the east end of the centre part of the city. I saw a thought-provoking sight as I strolled outside the Tower. Coming from the north and heading across Tower Bridge was an endless stream of army vehicles. Some of them were big trucks or lorries, carrying flattened out barrage balloons. Others were loaded with anti-aircraft guns and attendant paraphernalia. There were soldiers and airmen hanging on everywhere, some of them women. Apparently the air defences were moving south. I visited the roofless church of All Hallows Berking, across the road from the Tower. Although badly blitzed, its squat brick tower still stood. Samuel Pepys recorded watching the First Great Fire of London from there in 1666. The bombing had uncovered a doorway and a large cross in the walls, previously unknown. These originated in Saxon times and were at that point the only Saxon relics found in London.

I was standing all alone in sunshine in the middle of the church when a cluster of three or four flying bombs came across the sky from the southeast. Puffs of smoke showed the explosions of anti-aircraft shells behind them. (Mrs Churchill later recorded that her daughter's anti-aircraft battery in Hyde Park was one of those firing.) I moved under an archway as a precaution in case they came my way. They vanished into the distance to the south-west and I heard a series of dull booms. In a day or two I found that one had caused a heartbreaking tragedy. It hit the Guards' Chapel at Wellington Barracks, across the road from Buckingham Palace. A special church service was in progress, attended largely by the families of guardsmen who were on duty at the palace, together with a number of distinguished senior officers. The famous ceremony of the Changing of the Guard had just taken place and the Old Guard had returned from the palace. The men were standing in formation beside the chapel when the bomb hit. The nave collapsed, killing 119 members of the congregation and maiming dozens of others. The guardsmen instantly flattened themselves on the ground as the explosion took place, but immediately rose to their feet and stood at attention for a few seconds until they were ordered to dig out any survivors.

That afternoon I wandered around Hyde Park. This is the place famous for outdoor speakers urging everything from abolishing taxes to committing mass suicide. It was a lovely June day, the temperature just right, and the Londoners were taking full advantage of their day off. Families were picnicking on the lawns and there was the inevitable necking going on between couples lying on the grass, the men in service uniforms of all countries, the women in civilian dress. (The dowagers who sternly ran the women's armed forces took a dim view of that sort of thing, so the uniformed women were all upright.)

I strolled along in what seemed to be an idyllic peacetime situation until we all heard the now well known sound of a doodlebug engine approaching. As I observed the tranquil scene, the speakers continued to harangue, the children continued to play, the neckers continued to neck, the strollers continued to stroll, and the only recognition of the enemy onslaught was that a few people moved out from under the trees to see if they could spot this new-fangled gadget. It passed out of our area and a distant boom indicated that it was no concern of ours. I so wished that Hitler and Goebbels could have seen their panic-stricken enemy!

It was at about this time that I realized that the anti-aircraft firing had stopped. This and the movement of guns and balloons I had seen earlier were accounted for by a fundamental change in defence tactics. The authorities realized that it was useless to fire at a flying bomb and bring it down in one district when, left to its own devices, it would go on to another without selecting any special target. Also, our gunners were hitting very few of them as they flew at speeds close to four hundred miles per hour and were an impossibly difficult target. So on very short notice a policy that eventually proved highly successful was adopted.

All the balloons from London were moved to form a dense arc extending from the east, south of the suburbs, to East Grinstead. The barrage was increased in number during the summer so that by the middle of August, 2,015 balloons with dangling cables were set to entice the doodlebugs into their webs. All available anti-aircraft guns were placed south of the balloons or on the coast, leaving a strip in between and one along the Channel for the fighters of the Royal Air Force.

Our pilots employed various tactics. If they were close to the intruders, they could actually catch up to them despite their speed and explode them in mid-air with machine-gun or cannon fire. However, the pilots had to be careful not to be blown up themselves. Hence they developed a technique of flying alongside the doodlebug with the fighter's wing just underneath that of the V-1. The slipstream then flipped it over so that it lost stability and crashed to the ground in open country. Or rather, what it was *hoped* was open country, but obviously could well be a village or a country house. Whether this was considered a good idea or not depended on whether the considerer lived in London or the fighter belt, dubbed "Bomb Alley." However, it saved many lives. Postwar statistics indicated that the casualty rate per flying bomb in the London area was twenty-two times that of Bomb Alley.

However, life in the southeastern counties was still dangerous and nerve-racking. An added risk came from stray bullets fired by the fighters. In the issue of 23 June 1944, the *Sevenoaks Chronicle* published a vivid account of life in Kent, England's southeastern-most county. It reads as follows:

Early on Saturday evening [17 June] a man was getting into a bus when a bullet struck him above the wrist. Another man at his place of employment stopped a bullet in his leg …

At a Police Station bullets went through the window into two rooms which were unoccupied at the time. Bullets also went through windows of a bank nearby. Others pierced the glass door of a café and shop window without causing any casualties. The National Anthem was being played at an open-air gathering, but despite machine gun fire no one moved until the anthem was concluded. Then they took temporary shelter for a few moments, only to return for the continuation of the programme which actually went on without interruption.

… On Monday evening [19 June] one of the robots was brought down in close proximity to a large house situated in a pleasant wooded part of the countryside. A short burst from fighters sent it crashing into a tree, the explosion ripping away every branch and leaving only a gaunt and blackened trunk. The owner was walking in his garden, but when he heard the bomb descending he made a dive for a rhododendron bush. It saved him from injury, but his home was badly blasted … The verandah was shattered, and the furniture in every room turned upside down, windows were gone, and there were great fissures in the walls. All the ceilings had collapsed and furniture was partly buried under rubble.

On 7 July, the same newspaper printed further details of incidents in Bomb Alley:

The tranquillity of the countryside in Southern England continues to be disturbed by the shattering noise of German flying bombs …

Babies were killed when a London County Council Children's Hostel in Southern England was heavily damaged by a flying bomb in the early hours of Friday morning. It is thought that the [pilotless plane] was first hit by anti-aircraft fire before it fell in flames and exploded on the building, causing fatal casualties.

The house – a large county mansion – contained a staff of 13 adults and 26 children [evacuated from London].

Scores of rescue workers toiled with pick and shovel among the debris, but up to a late hour on Friday night the bodies of two children and three adults had still to be recovered.

The Head of the local Police said that he and five of his men were first on the scene, adding that the building was then a mass of flames. Before the work of rescue could be undertaken the men of the National Fire Service fought strenuously to get the flames under control. One by one the little victims were recovered, their identification being established by small labels attached to their ankles. One of the babies rescued alive appeared to be on the point of collapse, but it was then discovered that its mouth was filled with pieces of plaster. When this was removed the infant was soon sitting up and taking nourishment. The Matron of the hostel was among the saved but she was badly injured.

The mortal remains of the victims were laid to rest in a communal grave following a service in the Parish Church.

In a brilliant and complex operation in mid-July 1944, the guns in front of the balloons joined those already on the coast, thus widening the area left to the twenty-three fighter squadrons lurking there. A total of 1,596 guns and rockets were now aiming out to sea. Half the crews were women.

By then the artillery had obtained an American-made proximity fuse that caused the shells to explode when near the targets without actually having to hit them. This greatly improved the effectiveness of the guns.

Meanwhile, the Allies were bombing the launching sites whenever they could be found, and the armies were making slow progress eastward along the French coast. The bombers were also attacking V-bomb factories, transportation facilities, and storage dumps. All in all, a considerable diversion was being made from the march towards Germany.

These measures were generally known, but one of our most important defensive tactics was not disclosed until thirty or so years later.

In *The Double Cross System*, Masterson tells how a number of German spies, referred to in chapter 1 in connection with the invasion deception, were induced to send reports of both the V-1s and their successors, the V-2s, or rockets. Although correctly reporting the landing times, they gave false positions indicating that the bombs had reached farther west than was actually the case. This caused the Germans to shorten the range, thus gradually moving the attack eastward towards open country. As a result, the western part of London had very little trouble from the V-1s or V-2s, but it was hard on the easterners.

All these measures resulted in a great increase in the number of flying bombs destroyed before reaching London, but a long and difficult summer still lay ahead before the attack was reduced in scope. Then came the winter, with its V-1s launched from aircraft and its V-2 rockets. But all that was yet to come.

CHAPTER ELEVEN

"Just Carry On ..."

When we returned to our classes on Monday, 19 June, it seemed as though we had been away for months. Everyone had stories to tell and theories to expound. However, we settled down, plotted our observations, and drew up our practice forecasts as usual. There was an occasional alert and we heard the odd buzz bomb hurrying about its evil business. Otherwise life continued more or less as before.

Later statistics indicate that 750 civilians were killed and 2,700 seriously injured during the first week of the attack. Millions more had been under considerable tension day and night.

As I look back on the psychology of the whole experience, there seem to have been two distinct phases. The first few days, and even weeks, we were elated and buoyant in spirit. The much-boasted-about Vengeance weapon was having little effect. The very novelty of the situation gave the tired Londoners a lift. A return to conventional raids would probably have had a much more severe impact on morale, as was the case in the Little Blitz.

We were further buoyed by the news that the armies in Normandy were making good progress, except around Caen. Moreover, we could hear at night the heavy bombers heading towards Germany, and sometimes in the daytime the American Flying Fortresses could be seen taking over the day shift on the same deadly run, tiny silver crosses against the blue.

We went about our affairs, with everybody determined not to let Hitler's boast come true. Factories and meteorology courses continued. People were saying to themselves, in the words of the policeman, "Just carry on ..."

There were frequent alerts during the day, and, as before, we sometimes heard explosions, the sound of an engine, or often nothing, on which occasions the All Clear told us that the deadly invader had done its work somewhere else.

One thing I noticed within the first few days was that everywhere you went, people had circles under their eyes. It was an extraordinary unifying phenomenon – the fact that an entire vast city had had a series of poor nights. We all were sharing in the danger together, although at that early stage it seemed more a nuisance than a real risk. However, little by little the situation changed. This can be illustrated by two trips I made to hear a symphony orchestra at the famous series of concerts called the Proms, held in that monstrosity of Victorian architecture, the Royal Albert Hall. This big round concert hall had a domed glass ceiling under which a horizontal canvas canopy had been hung as a blackout curtain. The hall was filled with its usual quota of eight thousand music lovers, the majority of whom stood squashed together on the ground floor where there were no seats, the rest ensconced in boxes and galleries extending upwards to a considerable height. I sat in one of these and, with my chin on the rail, could see and hear everything very well.

During the symphony a little red light came on in front of the conductor. At the conclusion of that movement he paused, and a gentlemanly voice made the following announcement in impeccable BBC English: "Ladies and gentlemen, may I please have your attention. An alert has just been sounded. There will now be a brief pause, during which those who wish to do so may take shelter either in the corridors of this building or in the trenches across the road. The concert will then continue."*

---

* In Kensington Gardens, across the road, a number of trenches had been hurriedly dug at the time of the Munich crisis in the fall of 1938, when Britain found itself faced with the immediate prospect of a war for which it was grossly unprepared. These trenches were for the population to lie down in during air raids and would have given a certain protection against horizontal blast. Six years of neglect had done them no good, and their shallow bottoms were covered with water, mud, dead leaves, garbage, and the odd used condom.

At the mention of the trenches a titter ran through the audience, followed by respectful silence. The conductor stood motionless, and I watched the entire audience and orchestra. Nobody moved – not a soul. A bomb within half a mile would have caused that huge glass roof to come down in millions of sharp pieces through the canvas curtain and cut the standees to shreds. After a few moments the conductor raised his baton and the concert resumed.

As time passed a certain change gradually came over us as the enemy chipped away at the city. One of the first signs occurred when we went to a concert two weeks later. Several of us charged out at high speed after our last class of the day and went into London to hear a particularly attractive Proms program. We made some fast transfers of trains, buses, etc., and walked briskly towards the Albert Hall in the evening sunshine. We were surprised at meeting little groups of people coming away from it. When we reached the entrance there was a small notice on the door that read: "The management regrets to announce that due to the current situation there will be no Promenade Concerts until further notice. For obvious reasons, there could be no public announcement to this effect. We deeply apologize for any inconvenience which this may have caused."

We were disappointed and rather concerned. No one in our group could remember any previous occasion on which the Proms had been cancelled and the implication was that things were getting a little bit difficult.

Other evidence of the slow but steady deterioration in our situation began to show up. Fifteen of our twenty-five course members were quartered in different rooming houses scattered around the district. One morning after a rather noisy night my three fellow Canadians turned up slightly late for breakfast, carrying suitcases, their navy blue uniforms white with dust and their faces black with soot. A flying bomb had landed in the street and destroyed their rooming house, where we had all heard the D-Day broadcasts. No injuries resulted but it had been a nasty experience. They were given temporary cots in the basement of the old college.

A few days later another pair of classmates showed up, also covered with dust, and were also given quarters in the basement. There were more to come.

Two or three weeks went by and there was no let up. Our first euphoria passed, and although we maintained an outward cheeriness I think that most of us felt that things had gone far enough. They were to go farther.

During the first two weeks of the battle the average number of bombs launched per day was ninety-seven, sixty-five percent of which reached Greater London. The worst day of all was Sunday, 2 July, when 161 bombs crossed the coast. Reports of damage to famous places started to circulate. In the heart of London there is a short curved street called Aldwych where, on Friday, 30 June, a V-1 landed at a busy time close to a bus. Forty-five people were killed and 150 injured. A day or two later I passed by. The bus was lying on its side, completely flattened as though run over by a giant steamroller. Fortunately, there was no trace of the many fatalities.

An early bomb landed near the Marble Arch. Another hit the top floor of the Regent Palace Hotel, just off Picadilly, leaving the lower floors undamaged but killing a waitress. I had occasion to enter the lobby the next day. Things were proceeding normally on the ground floor despite the wrecked top story. One of our Wren officers, Paulina Brandt, reported that a bomb had hit her home village of Caterham, which had had a rough war for four years already. When she was an adolescent in 1940, she and her brother had been machine-gunned by a low-flying aircraft while crossing a field, fortunately without injury. The village had suffered through the Blitz and had sustained nasty casualties from the butterfly bombs in the spring.

On 3 July, a bomb landed in Chelsea, killing sixty-four soldiers, mostly Americans.

The defences were managing to shoot down a considerable percentage of the attackers or to stop them with the balloon barrage, but there was still a steady trickle getting through. In addition to homes, businesses, and factories, an average of one historic building a day was being destroyed. This was particularly hard on the morale of the Londoners, so justifiably proud of their heritage.

Every so often one would see sooty-faced people making their way home on a bus or the Underground, obviously having had a bomb land near them at their place of work. Our course members who lived in western London went home every night wondering what they would find when they entered their street. There were tragic stories of people who returned only to find their houses demolished and their families dead. This was quite different from the Blitz. Except for the few early daylight attacks, people could count on being home together during the air raids, as they mostly occurred at night.

I always felt that we Canadians and our neighbouring allies, the Americans, were most fortunate that our near and dear ones had the Atlantic Ocean between them and these terrible weapons. I was very glad not to be trying to analyze air masses or work out resultant winds while wondering whether that flying bomb that I could hear in the distance was heading towards my wife or my mother. I don't know how those fellows had the nerve to go home each evening.

For the first few nights after the main attack I went to bed in my room, going to the cellar only if things were getting close. Our nearby guns had moved out of town along with their colleagues, so that things were a good deal quieter. However, we soon decided that it was better for me to start the night downstairs on my two easy chairs rather than disturb the ladies later. There was no such thing as an attack-free night.

In the middle of the room was a stout table. I told my "room-mates" that whenever we heard the buzz of a bomb approaching we should duck under the table as there was space for all of us. The other inhabitants of our cellar room were the fuzzy dog and the two cats. George the Second and John the longshoreman found couches on the main floor. Nobody was left on the upper two levels.

As during the Blitz, many Londoners took shelter every night. Some houses had reinforced basements; other people slept with the spiders in garden shelters. Much publicity was given to those who spent every night in the Underground stations. They made good copy for journalists, and the informal entertainment they produced was easy to broadcast. But it should be emphasized that on the whole their numbers were grossly exaggerated in the news media. Only a small percentage of the population could possibly be accommodated in the tubes, and millions who lived in the metropolitan area were miles from any Underground station.

Also, as during the Blitz, vast numbers of English people preferred to ignore the alerts and live their normal lives. A story that received widespread circulation concerned a lady who decided to take a bath during a raid. A bomb landed nearby, tearing the tub loose from its plumbing and collapsing the house. The tub, complete with water and occupant, slid down the rubble and out into the street where it coasted to a splashy halt for all to see. As unfortunately no photographic verification was made at the time, the truth of this report cannot be guaranteed, but thousands of *raconteurs* swore to its authenticity.

After a week or so, it was obvious that the government was a little concerned. We now know, of course, that they were *very* concerned, but at the time there was no indication of this, nor that they were also highly alarmed about the forthcoming V-2s. Statements began to appear telling the public that when they heard the drone of an approaching pilotless aircraft, they should take cover immediately. The population was well used to dodging into doorways or under tables, stairs, or archways, but they had not been doing so to any great extent for these new creatures. Even the Admiralty took notice and issued an order that all naval personnel were to take extreme precautions against becoming casualties. We were ordered to seek cover whenever a flying bomb approached. Their lordships were becoming a trifle worried about the cavalier attitude and the general feeling of "it can't happen to me."

Something of the British character could be seen in an incident that occurred during lunch in the Painted Hall at the college. I have previously mentioned the stout oak tables. The order had just come out about taking cover. With a ceiling high above us the heavy tables would have given us considerable protection if it had decided to come down. Part way through lunch a buzz bomb could be heard approaching. We went on with our conversations, pretending not to notice but keeping one ear cocked to the south. This time the bomb didn't head away from us but continued its crescendo to a good loud roar. Nobody paid any attention until the sound stopped. Immediately I ducked under the table, and as I looked up and down the long row of human legs, I saw three other officers, all of them Canadian, dive under the table like myself. There was a loud bang, but the bomb had obviously struck across the river in Poplar, too far away to cause us any damage. I emerged to find that all the English members of the course had continued to eat as though nothing had happened. We four Canadians dusted ourselves off sheepishly and resumed our lunch. We were carrying out orders and doing the sensible thing. It seemed silly to sit there eating one's soup with an eighteenth-century plaster ceiling coming down all around and bashing in one's skull. However, the English, men and women alike, felt that it would never do to give in to a mere buzz bomb. To this day I don't know whether to feel ashamed of us North Americans or to consider that the English were being bloody stupid. But from then on we didn't duck under tables in the Great Hall.

I grew better acquainted with Miss Hards and Billie as a result of our shared nights. It became evident that Miss Hards in particular, who was the more talkative of the two as well as the boss of the establishment, had her own private attitude about the war. It wasn't so much a great conflict between the Allies and the Axis powers, or even between Great Britain and Germany. It was between her and 'im. She referred every so often to 'itler, but more often just called him 'im. She obviously considered that the rest of the war was of minor importance. After her favourite set of china was smashed she said to me, "I wouldn't let 'im mike me cry, not even when 'e broke me mother's best china."

I witnessed another example of British phlegm on a sunlit Sunday evening a week or two later. I had just returned to Greenwich when a flying bomb came over and exploded a few blocks away from the railway station. I could hear the fire brigades approaching the area. I walked towards it and found an old factory starting to burn fiercely. It was the Merryweather Fire Engine Works, reputed to be Britain's sole source of fire engines – much-needed equipment in wartime. The natives of Greenwich later told me that all through the war the Germans had been trying to hit that factory without success. But here, on this peaceful Sunday evening, a random shot had succeeded where countless bomb aimers had failed.

An interested crowd of spectators soon formed to watch the fire. Flames crackled upward, twice the height of the building. Black smoke poured out in great billows against the pale pink sky. Men, women, and children crowded around, commenting on the fire and chatting amiably about this or that. Children played tag around our legs. It was quite a sociable affair.

As dusk gathered and the fire raged on, an army captain rushed up to me and said in a desperate manner, "I say old chap, have you seen my smoke machine? I left it here a few minutes ago and I can't find the bloody thing!" I assured him that I hadn't, but would keep an eye out for it. A smoke machine resembled a large furnace, complete with chimney, mounted on wheels. They were used to lay down smoke screens to prevent enemy aircraft from observing targets. It wasn't needed here as the factory was busily making a tremendous smoke screen of its own. I never did find the smoke machine, and I only hope its owner retrieved it before he had to account for its disappearance to his commanding officer.

The whole scene was completely surrealistic – England's source of fire engines blazing away, the alert still on, the likelihood of another flying bomb killing dozens of spectators, and the pleasant, chatty mood of the crowd. It was the most incongruous scene I have ever witnessed.

Years later, I read that Churchill, on visiting Stalin, had said out of a blue sky, "Do you know why bulldogs' noses turn up?"

"No. Why?" said Stalin.

"So they can breathe and still hang on. They never let go."

CHAPTER TWELVE

Closer and Closer

I became fascinated with the battle I was privileged to watch. I truly felt that this was one of the great moments of world history. The residents of the United Kingdom had withstood the Blitz and the Little Blitz, but now had a new problem on their hands. The armies in France were trying to capture the launching sites of the diabolical flying bombs, but were not making the progress along the coast that they would have liked. Dozens of V-1s were being launched every day as July progressed. I visited quite a few places where I knew bombs had fallen and became something of an amateur expert on the damage they caused. When they hit a massive building on the roof, one or two upper floors were wrecked but the building usually remained standing. But when they hit among the row houses of the eastern or southern suburbs they caused damage of a magnitude not seen since the parachute mines of the Blitz. If a V-1 struck in the middle of the block, about two or three houses were instantly turned into a pile of rubble three or four feet high. Several houses on either side would be gutted but the walls would remain standing. Staircases would dangle precariously in the air and the roofs were blown off. A few houses on either side beyond would lose much plaster, window glass, and roof tiling, and the structure would be weakened. Any glass still remaining in the windows of the other houses in the block would be sent flying. Fortunately, unlike during the Blitz, no incendiaries were dropped, and there were very few fires caused by doodlebugs – except, of course, the spectacular blaze at Merryweather's fire-engine factory.

During these trying months the civil-defence services were magnificent. Thousands of men and women formed the vast network that attended to the wounded and bereaved, recorded each incident along with damage and casualties, rescued those buried in the rubble, or found homes for the homeless. The air-raid wardens were in direct charge of their districts. They were assisted by the heavy-rescue squads, which ran large machines such as cranes, the light-rescue workers, many of whom were women, the police, and the fire brigades, most of whose calls were for rescue work. In addition there were the medical teams, from first-aid and ambulance crews to hospital staffs. Tired, apprehensive, and exposed to danger, they all carried out their responsibilities with courage and devotion. Fortunately, there was far less rain than usual in southeastern England during the three summer months, so people were spared this added nuisance.

The statistics on deaths show that, whereas in World War I it took on average one hundred pounds of German bombs to kill one person, with the V-1s and V-2s it took two thousand pounds – inflation of a grim sort.

The fatality rate was nothing like the terrible battles of World War I, where the British army, for example, sustained sixty thousand casualties in one day at the Somme, but the raids were causing much damage. Railway lines were knocked out, commuting was delayed, and we were all short of sleep. However, the general cheeriness continued.

Greenwich was receiving about one hit every two or three days. This borough, covering roughly one square mile, was really getting more than its share. Croydon, the suburb to the south of London that had been the target of some of the earliest attacks in the war, was also having an exceptionally bad time.

I noticed that while we Allies were all united in spirit and London stood facing the foe as an integrated whole, nevertheless we Greenwichers breathed more easily when that snarling crescendo turned into a diminuendo followed by a distant boom. It wasn't that we really wished any bad luck on Poplar to the north or New Cross to the west, but we were relieved that they were the unlucky ones and we weren't – a disquieting sentiment.

I also found as the months went by that we all took a personal pride in our own bombs and were a little annoyed if we met someone who thought that their bomb was bigger than ours. We

Greenwichers in particular didn't like sob stories from boroughs that had a lower bomb count than ours. We were proud of our bombs and didn't want to have any lip from lesser places.

We became expert at hearing that distant droning sound and pretending we didn't. We also managed to route ourselves into places of comparative safety by the time the noise reached its loudest. Once I was walking along a quiet street, the only other people in sight being a couple of children playing on the sidewalk and an elderly lady left over from the Victorian age rustling along in ankle-length black taffeta. A doodlebug decided to test us. We could hear its drone in the distance, apparently heading our way. As the sound became louder and louder I had a strong impulse to run and hide under a doorway, but I could see the children continuing to play and the little old lady continuing her leisurely walk. I realized that they were aware that I, an officer in the uniform of His Majesty's navy, was apparently paying no attention to the approaching danger. Therefore I felt that the tradition of centuries impelled me to maintain course and speed. Soon some other unfortunate district had a mighty bang. I suppose my superficial composure gave them courage, but it was really only a reflection of their own.

One of my favourite activities at this point was to be heading somewhere in downtown London that required me to use one of the deeper Underground lines. Some had two long escalators deep in the bowels of the city. It was wonderful feeling safe for a few minutes before emerging into the street again. I suppose everybody felt like that, but on the surface we all looked completely normal except for the circles under our eyes.

Early in July the ban on unnecessary travel was lifted. My fellow Canadians and I took advantage of this opportunity to visit Oxford the next weekend, 7 July. We sat on the floor of the crowded train all the way. Unbeknownst to me, my cousin, Commander (later Rear Admiral) Jeffry V. Brock RCNVR, was also in Oxford that weekend but we never met. He was on a short leave from sea duty prior to taking up a new appointment as Spare Escort Group Commander Western Approaches. We both thought the other was thousands of miles away.

Our tranquil two days in that gracious university town contrasted sharply with the situation in London, which was completing its worst week of the battle; 820 V-1s had been plotted for that week alone. Since 15 June the Germans had launched 2,754 V-1s, killing 2,752 men, women, and children.

When we returned we found that the air-raid siren at the southeast corner of the college had received a severe jolt and now suffered from a bad case of laryngitis. Another doodlebug had landed near a shed beside the north-east corner of the grounds.

Being pecked away at in this leisurely fashion got on our nerves. When would it stop? The armies were making only slow progress in capturing the launching sites, and while great things were being accomplished by our defenders, there were so many attackers that they could not all be stopped.

We didn't know it at the time, but Sunday, 9 July, marked a new step in the story. Aware that the Canadian army was moving eastward along the French coast, the Germans started flying Heinkel bombers from Holland, each with a doodlebug suspended under a wing. When they were well out over the North Sea, London came within range of the V-1 and it was sent on its way under its own power. This meant moving guns and fighters to the East Anglian coast. The RAF fighters found it considerably easier to chase the slow Heinkels than the fast V-1s. Many a lumbering mother was shot down, although usually after she had dispatched her lethal little demon. These attacks increased as the armies swept eastward and were a considerable nuisance during the winter.

Perhaps it was that weekend that I had a Sunday dinner in the dingy restaurant at Waterloo Station. I shared a table with a beautiful woman of my own age, a captain in the British women's army, attached to the Royal Artillery. She was in command of a battery that fired night and day at the incoming flying bombs. This was the first break she had had since they moved the guns to the southeast in mid-June. She was a charming companion and I was enormously impressed and quite embarrassed to think that a woman was defending me while all I did was worry about dew points and barometric pressures.

After hurrying through our skimpy wartime meal I escorted her to her train and saluted her farewell. That was the last I saw of her, but fifty years later the memory is as clear as when it happened.

The Air Raid Precaution authorities called bomb explosions "incidents," a vague and stiff-upper-lippish term that combined clarity of meaning with understatement. By now in Greenwich there had been a flying-bomb "incident" every few blocks, including one or two on the high street. One day when I was at the college there was a bang from the direction of Ashburnham Place. When I returned to my room I found that the bomb had landed at the far end of the next street, Ashburnham Grove – that is, about two blocks from our house. It had caused the usual destruction of several houses and had broken some windows and cracked plaster in ours, two hundred yards away. Elderly Billie had been standing in front of the cellar window, probably trying to see the bomb. The explosion had showered her with most of the glass from the window, a few pieces still remaining in place. Fortunately, the explosion was sufficiently distant that the glass merely fell in on her and caused no injury. From then on I urged Billie to wear my helmet whenever raiders came close.

There were bulges and cracks in the plaster walls but no real structural damage to the house. Within a day a cheery little gentleman came to patch them up. He was Mr Baker, universally pronounced "Biker." I don't know his profession, but he brought over large sheets of paper and pasted them across the bulges and cracks to keep the rest of the walls from collapsing at the next bomb. We didn't have any means of patching up the cellar window and the nights in our room now were chilly, even though the weather was moderate.

The West End was still teeming with men and women in uniform as well as civilians. Theatres and nightclubs continued to attract many customers. For the first time, Americans in large numbers were sharing in the dangers of air attack. Whether a service person had an eight-hour or two-week leave, London was still the Mecca for those seeking entertainment.

The best day so far for the defenders was Thursday, 13 July, when they had to deal with only forty-two buzz bombs. Between 15 June, the first day of the main assault, and 15 July, out of the 2,754 V-1s that were launched, 2,579 arrived in England; 1,280 reached London. During the worst of the attack twenty thousand houses were being destroyed each day.

Without fanfare, the government had evacuated to safe areas 228,000 mothers, 537,000 children, and 53,000 aged, invalid, and blind persons to safe areas. However, there were still seven million people left in the city, all of them short of sleep, full of anecdotes, and highly annoyed at the Germans.

At the height of the battle, word came down the pipe that some admiral or other had decided that met. officers should know something about rockets. We were taken by train and bus through the largely undamaged western suburbs of London to the beautiful rolling green fields of Salisbury Plain. Here a group of soldiers showed us a low launching device that mounted a rocket about two feet long and three inches in diameter. We stood around for some time until finally it fired off with a very satisfactory *whoosh* and struck the ground a couple of hundred yards away. Having thus had a glimpse of the raw might of Allied firepower, we returned peacefully to London and our doodlebugs.

No further reference was made to rockets either in the course or at any other time in my professional career. Nevertheless, we were now indoctrinated.

Every so often another dust-covered classmate showed up for breakfast in the now familiar manner, but there were no injuries. By 14 July, eleven of my classmates had been forced to move into the college, leaving four of us in two different rooming houses. We had a feeling that things were closing in.

And so they were.

CHAPTER THIRTEEN

Our Private Bomb

On Saturday, 15 July, it was my turn once more to be duty officer. This meant hanging around the college and drawing up the day's weather maps to keep our sequence complete.

The college was virtually deserted, everyone having taken the opportunity to leave the area for a bit of a respite. After dinner I went into the small theatre where various concerts and entertainments took place. It contained a splendid grand piano and I played everything I could remember. Sometime earlier in London I had bought a copy of Debussy's *Clair de lune*, which I had always loved. I spent some time on it – so much so that even today when I hear its lovely strains it puts me back instantly to that strange, solitary, silent evening.

As dusk came on I went back to my room. Recently another course member had moved into our house, but there was no sign of him as he had gone home for the weekend.

I had made inquiries about travelling to Windsor the next morning to see the castle and was looking forward to the trip. I had also been invited to have tea the next day with the Reverend F. Donald Coggan, later to become the archbishop of Canterbury. My late father had been principal of Wycliffe College, University of Toronto, training young Anglicans for the ministry. Shortly before the war he had scoured England for a top-flight faculty member and had selected young Mr Coggan. It was an admirable choice, as Coggan's later career proved. He became an immediate success and a close family friend. He was at my father's deathbed in 1939 and participated in the funeral service. He had married my wife and me in 1942 in Wycliffe Chapel and recently permission had been granted for him to return to England to help reorganize his alma mater, which had been blitzed. (Transatlantic passages were only granted to persons of high priority.)

I don't remember when the usual bedtime alert was sounded – it was so common. By now the six of us always went to bed in the cellar sitting room (that is, three two-legged creatures and three four-legged). Billie had my helmet beside her bed, and I kept a flashlight beside mine where I could grab it and dive for protection under the table.

At about two A.M. we were awakened by the sound of a doodlebug snarling its way towards us. On previous occasions the crescendo turned into a most welcome diminuendo. But this time it simply grew louder and louder. I reached for my flashlight and sat up, *en route* to comparative safety under the table. The noise became so loud that the house trembled. Then, a sudden silence, followed a few seconds later by a strange swishing sound and a huge explosion just outside. I had not reached the table and I knew immediately I had been hurt. I felt sick all over, without any localized pain. The plaster from the old ceiling came tumbling down in chunks and two or three banged me on the head. I had dropped the flashlight and dimly saw its glow being buried by the falling rubble. We were now in total darkness and I could faintly hear Miss Hards and Billie asking if everyone was all right. I was pretty groggy and could feel blood running down my face and into my eyes. It was not long before I heard George the Second's voice in the hallway outside. He had immediately come down the stairs to help us and was carrying a large flashlight. He shone it on the two ladies and then on me, and I heard Miss Hards say, "Look! He's hurt." I tried to get up but my knees were wobbly and the rubble-covered floor was very uneven. With George the Second on one side of me and the two ladies on the other, they more or less boosted me towards the door a step at a time. His flashlight beam showed the air in the room to be dense white, and it was hard to breathe. At the foot of the stairs we met a burly soldier who had come from nowhere to help. He sized up the situation, picked me up, and practically ran up the stairs. I was dimly aware that "Grandfather" had fallen over and was partially blocking the hall. I remember my relief at getting out of that choking atmosphere into the cool night air, even though it too was filled with dust, the dust that was such a feature of these flying-bomb incidents.

The soldier set me down on the sidewalk leaning against an ornamental stone gatepost, and there I relaxed. A number of people were milling around and very soon an air-raid warden asked me how many residents lived in the house and whether they were all accounted for. My jaw was very stiff and I could hardly talk, but I did my best to list the people, totally forgetting the new officer (who was away for the weekend anyway). I was not in any pain: I just couldn't see, and I felt sick. In due course an ambulance arrived – not the peacetime single-passenger type, but the wartime version, with room for four patients, two above and two below. Emergency lighting had been rigged, the English having found by this time that it was much better to see what they were doing than to worry about pilots that didn't exist. I was put into one of the upper positions in the ambulance, and I could hear others being lifted aboard, one beside me and two below. Soon I could feel that the ambulance was moving. The voice of an elderly woman below me was lamenting the situation, and a younger voice beside me, who turned out to be her daughter, was comforting her mother. I joined in as well as I could and said, "We're all right now. They'll take good care of us," or some such inanity. The fourth passenger said nothing for the entire trip. I later found that he was the minister of the nearby Methodist church, and was dead.

We soon arrived at the Miller General Hospital, where the boy with the bleeding face had been taken a month earlier. I could see very little, but I soon realized that some kind nurse was washing my blackened feet. I remember thinking how horrified my mother would have been had she known. If, as a little boy, I had been playing in Manitoba's black soil and she found my feet dirty, she would say, "Oh, Son, what if you had been injured and had been taken to a hospital? I'd be mortified if anyone saw such dirty feet!" Here I was in a major medical emergency and the first treatment I received was a foot bath. When I told Mother this after the war, she *was* mortified.

In what seemed like minutes, they were preparing me for an operation. In fact, it was morning and I had been out cold under sedation for several hours. I dictated a message to Mr Coggan cancelling our tea engagement, but the lines were out and it never reached him. In the operating room one of the young doctors tried to inject the anaesthetic, pentothal, into my arm, but my veins had apparently collapsed and were nowhere to be found. The other doctor said, "Try his leg." Soon I felt a needle in the general neighbourhood of my ankle and was told to start counting at approximately the rate of one count per second. In subsequent operations where pentothal was injected into my arm I reached the number twenty or so before losing consciousness. This time, I kept counting to sixty or sixty-five. The other doctor said to the first, "Shove it in harder." I felt an intense pain in the vein of the inside of my leg and very soon sank into oblivion. (This procedure apparently caused my leg to develop varicose veins in a few months, as well as keeping me unconscious for about fifteen hours.)

After what seemed like only a few minutes I opened my eyes and it was dark – just a dim light near my bed. I asked where I was and a woman's English voice said, "At Queen Victoria Hospital, East Grinstead, Sussex. We specialize in plastic surgery." It was actually midnight. I felt fine but I could't see very well and could barely move my jaw, which apparently made me sound like Jimmy Stewart. My head and right arm were bandaged and I was told not to sit up. As I lay there mulling things over, I remembered reading an article in *Reader's Digest** some months earlier describing this place. It was famous for its work on badly disfigured faces, mostly those of airmen who crashed during the Battle of Britain. I decided that I must be pretty thoroughly carved up to have been taken there.

In a few minutes I could hear the well-known drone of a doodlebug passing overhead, which made me feel quite at home. The nurse, Joan Ricketts, a native of Surrey, was wonderfully helpful in reassuring me and administering to my principal need, which was thirst. I had not had any liquid in my mouth since leaving the ruined house and I was crunching plaster dust. I soon fell asleep.

I woke up the next morning to the first of a series of visitors. One was my uncle, Captain Eustace Brock, RCNVR, the senior officer of all Canadian naval types in the UK. He was one of the first to receive the casualty report and had come down from Greenock in Scotland to check up on me. He arrived just as they

---

* "The Town with an Educated Heart," November 1943, 100.

were changing my dressing. I had asked for a mirror and saw a terrible sight. My face was covered with blood and what seemed to be a thick yellow pus. As he arrived I pointed to my face and said, "Doodlebug bites," which I thought was rather bright at the time. He looked a trifle green. I soon found out that the yellow pus was not pus at all, but a new type of medication that had been held in reserve for use during the invasion. When the casualties in Normandy were not as great as had been anticipated it was released for use elsewhere, and I was one of the first in the UK to be treated with it. This, I learned, was called penicillin, and its early form was this terrible-looking jelly. Actually, I wasn't in too bad shape; my wounds being largely superficial. A self-inventory revealed a concussion and possibly a depressed fracture of the skull, a broken nose, a slight crack in my jawbone, a missing right eyebrow, and a number of cuts from flying glass on my face, ear, right hand, and arm.

All the glass was from the few fragments that were left in the window after most of it had come in over Billie a few days earlier. If the window had been intact when the closer explosion took place, I would have been turned into hamburger.

The fellow in the next bed was a British paratrooper. He had jumped into Normandy about ten minutes past midnight on D-Day, 6 June. As happened with so many others on that chaotic night he was dropped too far inland, receiving a nasty leg wound in the process. He limped around for ten days trying to evade capture and to find our forces. When he finally did so the medical officers put penicillin jelly on his wound. It was healing rapidly. They said it would have taken many weeks to cure if it had not been for the new wonder drug. In the days that followed we became good friends, although it was some time before I could turn my head far enough to see him.

I also was visited that morning by Dr (later Sir) Archibald McIndoe, pronounced MACKindoo, the head of the hospital, a brilliant pioneer in plastic surgery. As the day wore on I found more information about myself. No one remembered a pentothal anaesthetic lasting so long. I apparently now held the world record.

They had wanted to clear out all casualties from London before nightfall, and I was sent (still under anaesthetic) in an ambulance from Greenwich to East Grinstead, some thirty or forty miles. The only other passenger this time was the young woman who had been my ambulance-mate the previous night. When we arrived we were still in our bloodstained nightclothes. She was quite conscious, but of course I knew nothing.

I was put into the famous Ward 3, filled with battle-bitten fighter pilots, most with badly burned faces and hands. Immediately McIndoe visited me. The treatment I had been given put him into a rage the like of which eyewitnesses had never seen. A group of American plastic surgeons was visiting the hospital at that moment, and he sat me up and used me as a demonstration of what should *not* be done when giving emergency treatment. "He should have been wrapped up and sent to us right away," said McIndoe. Apparently they had removed a good deal of my skin, including my right eyebrow, and had done some pretty rough stitching – rough, at any rate, by plastic-surgery standards. Also, it was looked upon as horrendous that I was moved under anaesthetic, and as the hours passed and I remained unconscious they thought I was done for. Thus, they were very surprised when I came to at midnight and was quite chatty.

On Monday evening I met the senior nurse of the ward, "Sis" Mealey, a young red-headed professional nurse from Ireland. "Sis" was an appropriate name for two reasons. In the British armed forces a nurse is called a Nursing Sister and hence is addressed as "Sister" rather than "Nurse." Also, she became a sister to us all on immediate acquaintance. She could be heard going up and down the ward with her shrill voice berating the fellows for not taking their pills or not making their beds to her rigid standards, but underneath there was always a genuine concern and affection.

"Sis" was on duty that evening. I had dozed off when another doodlebug came over. As I woke up she dashed to my bed from the other end of the ward, threw her arms around me and shielded me with her body, as though to comfort as well as protect me. She obviously thought that I might have a Thing about flying bombs. Of course, I had heard my first East Grinstead buzz bomb the night before and was reindoctrinated, as it were, so I was quite all right. But I enjoyed the hug.

A day or two later another distinguished visitor was the Reverend Donald Coggan. He had been uneasy when I failed to show up for tea on Sunday, the more so the next day when my expected apology didn't arrive. He knew that things had been bad in the East End but could not get in touch with anyone. The official casualty telegram reached my wife in Cleveland stating that I had been seriously wounded by a flying bomb and was at East Grinstead. She phoned my mother in Toronto, who phoned Mrs Coggan (who was still in Canada), who cabled her husband, who thus received my alibi by a roundabout route. He immediately came to East Grinstead bringing his customary aura of warmth and good cheer.

I was under the immediate care of Wing Commander Ross Tilley, RCAF, another brilliant plastic surgeon from Canada who was learning the advanced skills developed by McIndoe.

Within a few days he operated on my nose, already a veteran of several incidents with hockey sticks and such, and I recovered slowly but surely. I kept complaining that I could barely move my jaw, but no fracture showed up on the X-ray. Nor was there any bruise on my face, just cuts and a bang on my head presumably from a brick that was once part of our garden wall. It is interesting to note that at the time there were some three dozen civilians in our hospital, all of them victims of doodlebugs, all with broken jaws, and none with any evidence of facial injury. It was thought that our jaws were injured simply by the wave of air pressure from the blast.

On Sunday morning my Canadian colleagues, hearing of my mishap, had gone to the house and collected my rubble-covered gear. The flashlight was still shining. Everything duly arrived at East Grinstead sometime later.

I wasn't allowed to sit up because of my concussion, and I proved to be no good on a bed pan. As the week wore on it appeared that my intestinal system had shut down forever. On Sunday, a week after my arrival, orders went out that I was to be given an enema. The appropriate expert showed up. She was an elderly nurse notorious for her enemas and feared by all. It was duly given, but nothing happened. As with the anaesthetic, they decided to give a second installment, and did so. Still nothing happened. "Well, wait a while and you'll be more relaxed," said the virtuoso.

The headquarters of the Royal Air Force took very good care of their disfigured airmen at East Grinstead and every so often sent pretty visitors to cheer them up. Sunday was the day when a delegation of gorgeous creatures in air-force uniforms came down from London. Several of them chatted with me, and the prettiest sat on the edge of my bed for an hour or so, talking about this and that. I was afraid that the enema might work, but it didn't, although every time I made the slightest movement I gurgled like several hot-water bottles. At about six P.M., our charming visitors went back to their flying bombs in London. Eventually, to the admiration and relief of all, the enema worked. A bucket brigade was instantly formed and soon all was well as another record was added to my laurels.

After a couple of weeks, during which I had been lying on my back with my head bandaged, who should arrive but Miss Hards and Billie. Both were very pink and moist, having walked a mile or so uphill in the heat. Miss Hards was carrying a large paper bag, perhaps eighteen inches high, filled to the brim with fresh raspberries. This was the most delicious fruit I had ever tasted. They were very rare in England at that time and enormously expensive. It must have cost her a month's allowance of whatever she lived on. The two of them greeted me with obvious affection and great apologies. They somehow felt it was their fault that I had been hurt in their house. (In the months ahead, I was touched by the number of English people who apologized for my having been bombed while in their country. It was an extraordinary example of their innate good manners.)

I shared the raspberries with my ward-mates, who were most grateful to Miss Hards. She apologized for not having visited me sooner, but she and Billie had cut their feet badly when helping me out of the room, walking on all the broken glass, as well as nails, and jagged chunks of plaster and china. It had taken all this time for their poor feet to heal. This was a very big trip for them – I think their first out of London since the war started five years previously.

Miss Hards could give me more information about our private bomb. The house was still standing, though uninhabitable. It had a bad structural crack or two and all the plaster was down, but the house might be salvageable. The local council had found them temporary accommodation with another family, who were ordered to take in these refugees.

It had been a particularly bad night. Another V-1 had landed in Greenwich Park in the deer enclosure. Once back to complete my course in the fall, I was disconcerted to find that they were still serving venison in the Great Hall.

Another flying bomb had crossed over the college grounds, narrowly missing one of the domes. It crashed into a row of houses just across the road from the western fence. This was in King William's Walk, where I had strolled on the eve of D-Day. The rescue workers turned up promptly and were carrying out their duties when a very rare tragedy occurred: another V-1 came across in exactly the same track. It, too, narrowly missed the dome and crashed into the remains of the houses, burying the rescue workers. The captain of the college ordered all the midshipmen to turn out and dig for any survivors.

When I was reported as a casualty on the Monday, the college authorities decided that enough was enough and ordered the two remaining met. officers to move out of their rooming houses into the college. Hitler was doing quite well as far as the met. course was concerned, leading thirteen to two.

One of my most vivid memories was the Great Ointment Incident. The summer was very warm, and lying on my back, never having really got rid of the mixture of soot and plaster that had been ground into my pores, I developed what can be inelegantly but accurately described as ball itch. I would wake up in the night practically tearing them off. When Dr Parks, the physician attached to our wing, made his rounds, I asked him about it, saying, "I know it's not VD because I've been a good boy." "Parkie" took a look and said, "Oh, that's some sort of ringworm due to the heat. As soon as you're well enough to be moved we'll get one of the orderlies to give you a nice warm soapy bath and then you can put some Wexworth's Ointment on the troubled area. That'll soon fix you up."

In a few days I was deemed well enough to be moved to one of the deep bathtubs used for burn victims, and I had a blissful soaking. I finally got rid of the last traces of 42 Ashburnham Place. Then I dried myself and applied the innocent looking transparent ointment they had given me in a jar. It immediately produced a most soothing sensation. What a relief! But after ten seconds it was like the explosion after a doodlebug's silence. Wow! The worst pain I have ever felt in my entire life! I thought I was on fire! I was practically sick as the agony doubled me up. Parkie's last words rang in my head: "That'll soon fix you up." Just as I decided I was turning into a eunuch the pain started to subside and in a minute or two had gone away completely, leaving me exhausted but happy and delighted to find that I was still a baritone. The itch never returned.

On many occasions in the years after the war I told this story to doctors, but none of them had ever heard of Wexworth's Ointment, and none of them believed my story. Then, some forty years later, my wife was bothered for several weeks with an itchy rash on one shoulder. No dermatologist had any success treating it. We were travelling in England, and as we went through Crewe I spotted a Boot's chemist's shop. I thought to myself, "I'll just see what can be done." When I asked the saleslady for Wexworth's Ointment, to my delight she went straight to a shelf and produced a jar, the sight of which evoked mixed memories. That night my wife applied it to her rash. Again, the preliminary soothing effect. Again, the explosion, the fire, the agony. Again, the pain vanished. And that was the end of the itch. Now at least one person believes my story. (The name of the ointment is fictitious, the events are not.)

In due course I was able to walk, though I had little stamina. One of my first trips was to the civilian ward where my ambulance-mate was still a patient. She was a good deal worse than I. All her teeth had been blown out, she had lost the sight of one eye, and her pretty face was badly scarred.

She had lived just around the corner from us in New Cross Road. Her husband was with the British army in North Africa. In their back garden was an Anderson shelter, partly underground, lying close to our garden wall. She had regularly been sharing this haven with her mother and the minister of the Methodist church next door. The bomb had landed at the entrance to their shelter, killing the clergyman and slightly wounding her mother. If the bomb had landed on our side of the garden wall our house would have collapsed on us and they would probably have been uninjured. As it was, the wall had absorbed the blast and had vanished, except for a few brick fragments found in my bed. Her home and the church had been destroyed.

The lady and I chatted for a while and then I returned to my ward. I never saw her again.

That is the true version. But the rumours! Everybody in the hospital knew that we had arrived in the same ambulance in our nightclothes, and everybody in the hospital assumed that we were lovers. I spent the next year trying to stamp out this gossip.

CHAPTER FOURTEEN

Guinea Pigs and Maggots

The air-force ward of the civilian hospital at East Grinstead was a marvellous place. Built for the terribly burned fighter pilots who had defended Britain in the fall of 1940, it had developed advanced techniques unknown prior to that time. McIndoe was not only a brilliant innovator and technician in plastic surgery but a great human being also. He insisted on several rules. In all other service hospitals the patients, whether army, navy, or air force, were required to wear bright blue pyjamalike garments wherever they went, so that they would not be served a drink in a pub. The New Zealander McIndoe fought the British service establishment over this, saying that we looked bad enough as it was without wearing those things. He said that as we had been disfigured while wearing the King's uniform, we should be allowed to wear it whenever we wanted, whether in the hospital or outside. But he had one stipulation: nobody could have a drink the night before he was to be operated on. The lads stuck to this rule rigidly.

McIndoe encouraged the townspeople to visit and improve our morale. They organized social events and took us for drives if they had the petrol. They also removed all mirrors from public places so that no patient would catch a glimpse of himself when he was out to have a good time.

A dreadful event had occurred on Friday, 9 July 1943. It was a rainy afternoon and many children who normally went for a swim after school in the outdoor village pool had instead gone to see an American western movie at a local cinema. At 5:17 P.M. a low-flying bomber attacked the centre of the village, dropping a number of bombs, circling twice, and then machine-gunning people in the street. One bomb landed on the theatre. Fire swept through many shops and offices. Rescue workers dug for bodies for twenty-four hours without rest. According to the East Grinstead *Courier* of 8 September 1989, the final total of death certificates issued was 123; 393 people were injured. This tragedy seemed to strengthen the bond of affection between the townspeople and the burned flyers.

The airmen were members of the world's most exclusive organization, the Guinea Pig Club. The name was a recognition of McIndoe's pioneering work in plastic surgery and burn treatment. To be a member an airman had to have been disfigured in a crash and operated on at East Grinstead. Most were ex-Battle of Britain fighter pilots – "The Few," in Churchill's words.

The nursing staff consisted of a nucleus of professionals augmented by a wonderful group of young women from the upper levels of society, now wearing the uniform of the VAD. The name of the organization for which the initials stood was virtually unknown. Intense recent sleuthing has disclosed the reason. Founded in the last century and famous for valiant service in several wars, this ancient and honourable institution was burdened with the highly forgettable name "Voluntary Aid Detachment."

With the manpower shortage being what it was, there were only one or two orderlies, and the VADs did much of the heavier and less-pleasant work as well as performing normal nursing duties. They worked twelve hours a day, six days a week, with breaks for meals, and then took it upon themselves in the evenings or on their infrequent days off to date the patients. Aware of our fear that women would never go out with us again, they invited us to dances, picnics, and other social events. We were not pleasant sights, but the fact that they would spend their time with us was an enormous boost to our morale.

Another McIndoe rule was that we were to be treated as if we all had the same rank – no division between officers and men. Some of us were learning to make beds for the first time in our privileged lives.

A British army captain my own age had arrived from Normandy with a horrible wound. A bullet had entered his neck, leaving a tiny hole. It went through his throat and out the other side taking half his jaw with it. He had recovered generally, but the surgeons had a great deal of work to do before he could keep food in his mouth, as it leaked out through the large hole. A very buoyant spirit, he disliked having to make his bed. He patented a superb technique whereby, having once made it to the exacting specifications required by the Matron, he could wriggle in, have a good night's sleep, and emerge without leaving a trace. This meant approaching it from the head end and slowly inserting himself between the sheets, being careful not to make any wrinkles. The manoeuver took about fifteen minutes. He trained himself not to turn over during the night and then in the morning slithered out with the same patience, skill, and determination displayed the night before. One quick smooth of the hand removed any trace that a human being had ever been near the bed. His technique was much admired and imitated by many, but none of us ever approached his consummate artistry.

After I had been in the hospital about two weeks a new wing was opened, paid for by the Canadian government primarily for Canadian airmen, but open to all Allied casualties who required plastic surgery. The *British Journal of Plastic Surgery** gives some interesting background information on this addition to the hospital:

The Royal Canadian Air Force had been playing a greater role in the war with a subsequent increase in casualties. The Canadian Plastic Surgery and Jaw Injuries Unit under the command of Squadron Leader Ross Tilley, surgeon, with an anaesthetist and three nursing sisters had been attached to the hospital for training. The Canadian authorities expressed a wish to erect a building as a memorial to the Royal Canadian Air Force crews who died in the war. In June 1943 the Canadian government granted 20,000 pounds towards the cost of the building ... On December 11 the foundation stone was laid ...

Nine patients were admitted to the Canadian Wing on July 12th, 1944 and in early August all forty-nine beds were occupied.

I was the first and, for awhile, the only bed patient in the wing. I had a private room, which enabled me to catch up on much-needed sleep as the long, low wooden shed known as Ward 3 was a pretty noisy place. All this time there had been intermittent

---

* Issue 41, 1988.

doodlebugs passing over our heads on the way to London. One night I was wakened by that old familiar sound, the droning snarl of a buzz bomb. But as it grew louder I could also hear a smoother purring that I knew to be that of a Rolls-Royce Merlin engine. One of our fighters was obviously in hot pursuit. I heard a rapid burst of firing followed immediately by a small explosion and tinkling from the next room. Our fighter had fired one desperate burst before he had to give up the chase as he approached the balloon barrage, which, it will be remembered, extended from East Grinstead eastward. Two or three little explosive shells from the airborne cannon had hurtled into the unoccupied room, smashing the brand-new wash basin and knocking a rung out of the iron bedstead. As the only naval type in the hospital I gave my new air-force friends a hard time about their marksmanship.

Soon afterwards I received a visit from my wife's brother-in-law, Captain John Hare of the Royal Canadian Army Medical Corps. A doctor, he had been serving with the First Canadian Division in Italy and had been moved to England *en route* to Belgium and Holland. He had heard about me in a letter from home. In chatting we discovered that he had visited Windsor Castle on the same day as my cancelled excursion. We might have run into each other on those royal premises if my bomb hadn't intervened.

Our conversation was interrupted at one point by a buzz bomb flying directly overhead. Johnnie had not been close to a V-1 so this was his first experience at ducking away from a window. His comments were vivid, colourful, imaginative, and quite unprintable.

I couldn't read or write, but in the early weeks I kept in touch with home through the kindness of various visitors who read me my mail or wrote letters that I dictated. I was quite concerned about how my wife would find my appearance when I was finally patched up. Having been scared at my own reflection when I thought that the penicillin was pus, I had a further false alarm. The Red Cross gave to each survivor an emergency kit – a little cloth bag containing toothbrush, toothpaste, etc., and a small shaving mirror. Because broken glass was such a menace the mirror was made of shiny metal. It was about the size of a postcard. Somehow I had been given two of these sets and I used them interchangeably. When I was well enough to sit up I looked at myself in the mirror. All my hair was shaved off. My head and forehead were covered with fancy needlework. I was quite a mess. But what was most disturbing was that the entire shape of my face and head seemed to have changed. I looked practically sub-human, with a very narrow face and head, more like something straight out of an egg. I felt quite depressed by this but in a day or two had another look in my little bedside mirror and I didn't seem so bad. The scars were still there but I looked like myself rather than some science-fiction creature. A few days later I checked again; once more I looked peculiar. Then by chance I noticed that the mirror was slightly convex. The effect was some-thing like the Hall of Mirrors in a midway. I compared the two mirrors and found that the other was perfectly flat. So much the price of vanity.

Two ubiquitous sights around the hospital were Lionel and Stanley. They were about six and eight years old (or vice versa). Time has blurred my memory of which was which, and even their names. But not their personalities.

Both had lost their parents and other relatives in air raids and were alone in the world. One had terrible scars on his face from picking up an incendiary bomb, while the other had had a hand blown off, probably from a butterfly bomb. They had met at the hospital, to which they had been shipped from East London for repairs.

Although theoretically based in the civilian ward they spent all their time in ours. They much preferred talking aircraft with the RAF types to listening to the civilians going on about their grand-children. Months of satisfying their curiosity had given them a formidable knowledge of everything that could fly or shoot.

No one seemed to be in charge of them, yet everyone kept an eye on what they were up to, which was generally mischief. "Stanley! Stop it!" and "Lionel! Cut it out!" were heard up and down the ward from dawn till dusk. They spoke the language of Eliza Doolittle, which meant that hardly anybody could under-stand a word they said.

On one celebrated occasion the younger one (probably Stanley) had been playing in the mud and had smeared some across his face. After a while one of the RAF types could stand it no longer and told him to go and wash it off. He went away and returned a few minutes later looking dirtier than ever. The smears had just moved around a bit. On being bawled out he said, "Bu' I 'ave washed me fyce!" His critic picked him up, carried him to the nearest wash basin, and set him down in front, where the little head just barely reached above the edge of the basin. The RAF type pointed to the soap and said, "Now get busy."

Stanley, realizing the inevitability of the situation, turned on the cold tap with his one hand. Looking like a man who was being forced to commit suicide, he cupped his hand under the tap and slowly moved it to his face. It looked very much as though this were the first time in his life he had ever touched water.

The rapidly assembling audience shouted encouragement. Stanley looked around at his tormentors like a dying doe.

"Go to it, Stanley. Atta boy."

After several more applications of the dreaded fluid his face was distinctly cleaner in places, so the committee overseeing the operation, feeling they had done as much as they could, excused him. He sprinted off at high speed, terrified lest anyone should suggest a bath.

McIndoe had another bright idea that we didn't all appreciate, although we should have. At most hospitals occupational therapy consisted of weaving baskets or performing other tasks that young men tend to feel is "sissy work." Knowing this, McIndoe persuaded a nearby aircraft-instrument factory to establish a small branch in the hospital. Two or three professionals supervised patients who worked at repairing aircraft instruments for an hourly wage. However, it was exacting, precise work and, while undoubtedly better than hooking rugs, nevertheless was not as much fun as sitting around doing nothing. Each morning all walking patients were liable to be ordered by the RCAF flight sergeant to work in the shop that day. I developed enormous skill at becoming invisible when he made his rounds, jerking his thumb backwards over his shoulder like a hockey referee giving a penalty. But one day my luck ran out and I was banished to the shop. I had to clean the horrid little pistons in turn-and-bank indicators. I found this nerve-racking, boring, and tedious. However, I survived and even got paid.

Another of McIndoe's fine ideas was much more successful. Understanding young men, he knew that they were fascinated by things mechanical. Therefore he discussed the techniques of plastic surgery openly with everyone. Moreover, he encouraged patients to come into the operating room and watch operations as they took place. As a result it removed most of the fear and mystery out of surgery. The patients knew exactly what was taking place, and some of the fellows who had been there for four years quite literally knew more about the subject than most general surgeons.

However, there was one operation we were forbidden to see if we were scheduled to have it ourselves – a nose refracture. For repairing crooked noses the technique was to take a rubber hammer and hit the unconscious patient's nose many times from a great height. The splintering sound was said to be sickening. When the bones in the nose had been smashed into innumerable little pieces the surgeon was able to mold it into a beautiful new shape. In a few days, during which the patient developed two very black eyes, the nose healed and was back to work greatly improved. I never risked witnessing this; I had had a refracture soon after I arrived and couldn't be sure I wouldn't have another!

During my several stays in the hospital I saw a few operations. One of the most interesting was when Wing Commander Tilley sewed a set of eyelids on a burned airman, using the patient's skin. When fully healed the results were not only medically satisfactory but were remarkably handsome as well. Prior to that the pilot had not been able to sleep well because his eyelids were so shrunk that it was not possible to close them. There was not only a problem with light, but he would occasionally be wakened by the blankets scratching his eyeballs. Plastic surgery at East Grinstead was not for glamour, but chiefly to enable people to live more comfortably.

Plastic surgery also includes care of the hands. There were many air crew in the hospital whose hands as well as faces had been badly burned. In a number of cases the poor lads themselves were partially to blame. The Royal Air Force had strict rules to the effect that leather flying helmets and gauntlets were to be worn in the air at all times. These were considered a nuisance by some flyers, but they afforded considerable protection from the terrible burns from gasoline fires that were such a common feature of crashes. The fellows who wore their helmets and goggles were badly burned around the nose and mouth, but the rest of their heads and hands were intact. The others had no ears and their hands were red stumps, sometimes with one or two claws. When

I was well enough after about a month to be up and around, I could see more of my fellow patients and the effect they had on visitors. People unfamiliar with burn wounds were sometimes almost sick at the sight of what war had done to these gallant defenders of freedom. However, as one became better acquainted with these fellows, they were usually found to be perfectly normal, healthy, reasonably happy young men who loved a bit of fun. Two or three were still rather morose, but the kindness of the doctors, nurses, and the townspeople was gradually helping them realize that they were still welcome and respected members of the human race.

One aspect of plastic surgery that we became used to seeing but that shocked visitors was the intermediate stage of a nose graft. Pilots who had had the fleshy part of their nose burnt off, leaving huge nostrils and a tiny bridge, were first given a fleshy nose from their chest. It was not disconnected at first. A tube of their own skin ran from the end of their nose down inside their uniform like an elephant's trunk. This was necessary to nourish the recently grafted flesh. A sign of improvement was that the tube also nourished itself and became fatter until it was about an inch thick. The plastic surgeon on his rounds would see this, give it an affectionate pat or squeeze, and say, "That's coming along just fine!" After a few weeks they would trim the nose, and it would end up looking almost normal except for tiny stitches and a bright red coating of mercurochrome. It was customary to leave the grafts open to the air, painted scarlet. Thus, visitors saw some fellows with really hideous faces, others with scarlet noses or eyelids, still others with a dangling trunk – all the more credit to the young women who chatted or danced with us and made us feel that we were acceptable companions.

One of the less charming features of East Grinstead was maggots. The treatment for certain types of severe burns was to place the affected area in a plaster cast. In due course maggots developed spontaneously under the cast, for reasons unknown to the patients. This was entirely approved of by the doctors, as the helpful little creatures ate away the burned flesh without harming anything else. However, they had one social shortcoming: they emitted a smell to end all smells. It permeated the cast, the room, and the corridor. Fortunately, the patient himself was incapable of detecting the odour. The doctors and nurses showed great dedication and self-control in their ability to enter the room and deal with the patient without wincing or throwing up. Many a wife or girlfriend sat all day beside the suffering patient without ever disclosing the secret. When the casts were removed the first thing the airman saw was a dense, writhing mass of thousands of beige-coloured worms, each about half an inch long. When scraped off, the relieved patient could admire his nice new pink skin that showed no trace of burn.

One grim event cast a cloud over the hospital for a few days. An American bomber returning from a raid crashed nearby. Two or three crew members survived, but they were badly burned. They were brought to East Grinstead before being transferred to an American hospital. One of them died while with us. Death was a very rare event in a plastic-surgery hospital. It produced a spate of such remarks as "Shockingly bad form"; "That simply isn't done here"; "Just like an American ... they always overdo it."

Let no one criticize this type of graveyard humour. Every airman in the hospital had been burned in a crash. Scoffing at death was how they stayed sane.

CHAPTER FIFTEEN

Flashback: From War to War

It was some time before I could read, due to bandages and lack of cooperation between my eyeballs. This gave me time to contemplate my career as a warrior. It was highly unimpressive.

I was born on 6 January 1918, in Winnipeg, which at that time boasted the proud title of being the world's coldest city of its size. World War I was raging. From time to time platoons of new recruits marched from Tuxedo Barracks past our house to the railway stations *en route* to the terrible trenches in France and Belgium. Whenever there was such a parade, the grownups took me out and waved my hand at the troops, who gave a cheery wave back. I sometimes feel that this was the greatest contribution I made to the Allied cause in two world wars.

We all grew up surrounded by the atmosphere of trench warfare. My father's younger brother, Lt A.D. McElheran, MC, lived with us after he came home until I was seven. He had been through many of the worst battles, was wounded three times, and twice buried alive by shell fire. His decoration testified to his bravery. He regaled us all with stories of the trenches, stories that I made him repeat endlessly. On my mother's side, Uncle Eustace commanded a motor launch in the Channel, Uncle Reg served with the British navy in the Mediterranean, which had little big-ship action, much to his disgust. Uncle Freer was loaned to the Royal Artillery. He rose to the rank of major and was given the command of a battery. He won the Distinguished Service Order for saving his guns at the time of the Big Retreat in 1918, when our forces lost one thousand big guns in a few days. Mother's youngest brother, Cecil (known as Tom), was a fighter pilot who was shot down three times and badly injured, but he managed to bag about a dozen Germans. Her sister Irma served as a VAD nurse in France. My grandmother moved to England and kept a house where her children and their friends were always welcome to spend a few days' leave from the front or the sea. I can remember her telling stories of Zeppelin raids, and how the chandelier over the dining-room table would sway with the explosions.

My father, rector of St Matthew's Church, was the Winnipeg chaplain to several battalions. More than 800 men from the parish served overseas; 112 were killed. Dad had been named as the next of kin for many of these men and received the grim telegrams. He had to bear the tragic news to mothers, wives, and sweethearts.

After the war, we frequently had services to which the uniformed survivors would parade and largely fill the nave. It was a stirring sight, and accounts of battles were read as part of the church service. There was a great atmosphere of respect for the fallen, far more so than after World War II. They were looked upon as semisaints.

The military hospital was full of disabled men. Many still wore a slightly greenish complexion from the first gas attacks at Ypres in 1915. The first gas that was used fell on Canadian troops, many of whom were from our parish.

We moved to Toronto in 1930 when Dad was appointed principal of Wycliffe College at the University of Toronto. The 1930s were a time of deep gloom for us, largely because of the Great Depression and the drought on the prairies, where so many of my father's former parishioners had settled. In Queen's Park, across from our house, there were always ragged, unemployed men sleeping on benches covered with old newspapers. Many were veterans, still proudly wearing their ex-serviceman's lapel badges.

But a new mood arose in which much of what we had earlier been trained to admire was denounced. Many were now strongly of the opinion that the servicemen had been suckers, victims of manipulation by the armament manufacturers. Nevertheless, in Toronto there were still spectacular parades. The drab khaki of wartime was replaced by pre-WWI dress uniforms, with red, gold, and glitter predominating. There were lots of horses: the cavalry was slow to acknowledge that a horse was no match for a tank.

We had cadets at school and spent many a frosty fall morning parading up and down Varsity Stadium. The idea that World War I had been a war to end wars had largely faded. Japan had invaded China and there were daily reports of the slaughter. In September 1932, as we started the fall term, we were given the choice whether or not to study German. I elected to do so on the naïve grounds that as the next war was coming and would be with Germany, I wanted to be a spy. That seemed much safer than fighting in the trenches.

Many of our generation were convinced there would soon be another war with Germany even though Hitler was not yet in power. However, our leaders didn't seem to be so certain and lagged far behind in their preparations. This illustrates a viewpoint held by people of my background about World War II when it eventually came. We didn't really look upon it as a new war, but rather as a continuation of World War I with the same enemy. Whether they were called Prussian militarists or Nazis, whether the leader was called the kaiser or the führer, we were still fighting Germans, as our fathers and uncles had. But pacifism was growing. An Englishman named Beverly Nichols wrote a book called *Cry Havoc* that had a tremendous influence on young men of my vintage in the late 1930s. The first part described the horrors of wwi, from trench warfare and the war at sea to the bombing of England. Then it described the tremendous advances in technology that had taken place since the end of that conflict, particularly in aviation. It also detailed the development of gas capability and the possibilities of germ warfare, as well as the greatly increased efficiency of explosives. It really scared the wits out of us. The book went on to ask, "Why fight? What if Germany did invade Britain? Would that really be terrible? Surely it would be better to let them come rather than go through another ghastly world war."

What Nichols never dreamt of was a nation that contained such unspeakable horrors as the Gestapo and the extermination ovens, a nation which would slaughter whole villages of civilians as reprisals for acts of sabotage.

When Mussolini invaded Ethiopia in 1936 a spirit of war arose, but when the League of Nations backed down it soon subsided and Mussolini went ahead.

In the late 1930s, when we went to the movies to be entertained, we often ended up depressed because of newsreels that showed enormous quantities of marching men and the build-up of military hardware in Germany and Italy. Our parades seemed so puny in comparison, with their handful of wwi field guns drawn by horses, and thin columns of men. But we cheered ourselves up by saying, "Stanley Baldwin [the British prime minister before Chamberlain] knows more than he says." He didn't. Or, "The German people won't put up with this again," and so on. We had heard stories of Hitler's police-state methods. I well remember a German ex-submarine commander who had become a Lutheran pastor. He preached a sermon in Toronto and told of the atmosphere of fear and suspicion that had spread through Germany. Friends were informing on friends, children on their parents. People who had expressed any criticism of the government, its economic or rearmament policies, or its harsh treatment of the Jews were very apt to be taken off in the middle of the night and thrown in a concentration camp. In many cases their relatives had no idea whether their loved ones were killed or merely imprisoned. This happened not only to Jews but to many German theologians, professors, and other intellectuals, as well as ordinary decent German people who didn't like the way the situation was developing. We said, "The Germans are certain to overthrow him soon." Unfortunately, as with most other dictators, things had been allowed to go on too long and opposition was impossible.

In March 1938, Hitler annexed Austria to the cheers of the German-speaking population. In the fall of 1938 the crisis over the Sudetenland held us all in a state of tension. This strip of Czechoslovakia had been part of Germany before wwi but was given to the newly created Czech state by the peace treaty of Saint-Germain in 1919. It was inhabited largely by Germans who were anxious to rejoin the Fatherland. To me and many others, it seemed like a reasonable request, and we were very relieved when British Prime Minister Chamberlain backed down at the Munich Conference and announced, "It is peace in our time." Those who criticized him often failed to realize that not only was Britain unprepared militarily for war but we who were destined to fight were unprepared emotionally.

However, in March 1939 Hitler seized the rest of Czechoslovakia despite his assurance to Chamberlain that if given the Sudetenland he would make no further territorial claims in

Europe. This action converted the British prime minister over-night from a friend to an enemy.

This still seemed rather remote to us, as did the fight that Hitler was picking with Poland. His propagandists fooled us into think-ing that the Poles were making hostile raids on Germany and killing Germans. After the war it was definitely proven that the dead German soldiers who were shown to the world press as proof of Polish aggression were actually inmates of German mental hospitals dressed up in German uniforms and then killed by their own countrymen.

One reason why we were not too concerned over Germany and Italy was the strength of the Soviet army. It had always been assumed that communists and fascists were natural enemies, like dogs and cats, and that if Hitler attacked his western neighbours the Russians would move in on his eastern front and soon finish him off. Thus it was a terrible shock when on 23 August 1939, a friendship treaty was signed between the Soviet Union and the Third Reich. This was the signal for war.

Early on Friday morning, 1 September 1939, we heard from various sources that the German army had crossed the Polish border at several points. While its army resisted the invasion, the Polish government asked Britain and France to honour their treaty obligations and come to its aid. Chamberlain sent a note to Hitler telling him to pull back or war would be declared. A deadline was set for 11:00 A.M. on Sunday. We spent two days hoping that the German people would rise up and denounce Hitler or that reason would prevail and he would withdraw his armies. But there was no such interference with his aggressive design.

We awoke on Sunday, 3 September, to find that while we were asleep the deadline had expired and Chamberlain had declared war. We went to church realizing that once more we were at war. It was like 1918, and all the stories we had heard when we were growing up came flooding back to our minds. I had often seen in the back of the prayer book the Prayer for Victory. For the first time it was used in the service. We also sang the National Anthem, which in Canada in those days was *God Save the King*.

We were all in a state of mixed emotions as we continued to listen to news broadcasts for the rest of the day. After years of wor-rying over the prospect of war, we were strangely exhilarated now that it was finally here. Having harboured latent guilt feelings about ignoring the Hitler menace, at last we were going to do something about it.

At about 10:00 P.M. we heard on the radio that the liner *Athenia*, bound for Canada, had been torpedoed without warning 250 miles west of Ireland. The broadcast said that the weather was calm and that there was a bright moon. Rescue ships were picking up survivors. Eventually we learned that 128 of the 1,418 passengers had drowned.

I went for a walk around the placid university campus in a state of shock. The moon was shining brightly. It was an uncanny feeling to realize that the same moon was shining on hundreds of people swimming around in life-jackets or shivering in lifeboats.

The invasion of Poland no longer seemed remote. We had been attacked too. The war was on.

CHAPTER SIXTEEN

Flashback: Conversion to Old Salts

The next morning brought two surprises: 1) Canada was not at War; 2) neither was the US. It appeared that with the new status of Canada as a nation we could only enter a war if Parliament declared it. Parliament, however, was not in session, but soon met; as of 10 September we were officially at war with Germany.

To our astonishment, the Americans did not see things as we did. We had felt very close to them and thought the two nations were much alike, but suddenly here was a totally different viewpoint. President Roosevelt spent the next two years proclaiming that his most important goal was to "keep America out of war." This was a political necessity as there was a great deal of isolationist and even pro-German sentiment in the country. A few, like Senator Claude Pepper, spoke on behalf of a cause that we felt should be supported by all freedom-loving democracies.

On Monday morning, 4 September, I had occasion to go downtown to a law office, as my father had died three weeks earlier of a coronary and we had legal matters to settle. I walked past the old armories. On the parade ground there were two or three recruiting officers seated at small tables. A few young men in civilian clothes were standing in line to join the army. Across the street was a considerable crowd of poorly dressed men, mostly unemployed middle-aged veterans of World War I. They were all trying to make up their minds whether to join up and have another go at it. One fellow boasted to his friend, "Why, I betcha a buck that if I went over there right now they'd take me."

Within a day or two the park across from our house was filled with the sound of "left, right, left, right" as sergeants and corporals trained the new soldiers, still clad in their shabby civilian clothes but wearing new armbands. As the weeks went by this continued all day long, and eventually they showed up in the new battle dress that none of us had seen before. But still they marched. It appeared that we were going to defeat the Germans by walking over them.

Canada was still in the throes of the Great Depression. The government urged all of us who were still being educated to continue with our studies until they could recruit large numbers from the masses of unemployed and had room for the rest of us. So I carried on with my musical training, but soon applied first to the Royal Marines as a bandmaster, and then to the Canadian navy, but was politely turned down by both.

During the summer of 1940 many English families sent their children to homes in North America to keep them safe from the air onslaught expected at any time. Complaints were made that this privilege was only available to the reasonably well off and that the poor were being neglected. Hence the British government, which was already evacuating tens of thousands of children from the cities to the countryside, commenced a program of sending children from poor families to Canada. They did so with some misgivings; it was obviously impossible to find shipping space or homes for the vast numbers of children living in the low-income areas, and there was always the possibility that a ship carrying children might be torpedoed.

In September 1940, I was unemployed, having been unable to find a job anywhere despite my BA. The people at the reception centre for British children in Toronto were at this time snowed under and asked for help. I applied and was taken on as a lifeguard and swimming instructor, my only qualifications being that I had free time and could swim.

We used the fine pool at Hart House, University of Toronto, the men's recreation centre located five yards from our house. Designed for water polo, the pool was six feet deep throughout, but was drained for our purposes to a uniform three feet. Dozens of naked boys were poured in. Bathing suits, considered an effete extravagance, were never worn in the Hart House pool.

It was probably the first time these kids had removed their clothes since they were born, and their bodies were white as snow. Dotted here and there amongst the squirming, screaming, splashing mass was an occasional tanned back that showed where my fellow lifeguards were vainly trying to maintain law and order.

None of these children could swim; nor, apparently, had they ever been in water before. They jumped and splashed and wrestled and tried to drown each other while we pulled them apart and kept an eye out for the littler ones, some of whom were at first terrified. But not for long. They showed an incredible affinity for the new element and were for the most part utterly fearless. Hitler was going to have a tough time if he tried to subjugate East London. In due course I was to see the same quality in their parents and those children who were left behind.

I invented a rough-and-ready method of teaching swimming that took about sixty seconds per boy and in two weeks taught many dozens how to stay afloat, much to their delight. However, their style was not of Olympic standard, resembling more that of a cross between a dog and a waterfowl. After two weeks the supply dried up as the evacuation came to an end. The British government's doubts about the scheme were justified when, on 17 September 1940, the refugee ship *City of Benares* was torpedoed in rough seas. Seventy-seven out of the ninety children on board were drowned, along with 171 adults. That ended the transoceanic evacuation and the influx of children ceased.

I had been a part-time demonstrator in geography at the University of Toronto and at the end of September managed to join the Canadian Department of Transport as a civilian meteorologist. Having spent the spring trying to conduct orchestras when my mind was in Norway or Dunkirk or flying over Kent, it was a relief to be doing something for the war, even though it consisted at first of just taking weather observations and drawing isobars. On completion of a winter-long intensive course at Toronto I was sent in the spring to the air station at Camp Borden, where fighter pilots were being trained. That fall I became senior met. o. – in fact the only met. o. – at Rockcliffe Air Station, just outside Ottawa. We had a number of new aircraft there and a good many others passed through on their way west. My most important task was when Churchill spent two days in Ottawa and I prepared special forecasts twice a day for flying his mail to and from Washington. My duties kept me at the station, however, so I was never able to see the great man.

On one interesting evening Ottawa decided to give itself a practice blackout. Two Civil Defence civilians were to fly over the city in a Norseman aircraft to observe the effect. I convinced them that their lives would be in serious jeopardy if they didn't take a weatherman along, so they let me hop in and up we went.

We flew over Ottawa at a height of several thousand feet and admired its normal brilliant lighting. Neon signs, street lights, windows – all shone like jewels. Then, at a set moment, the lights went out in large blocks or individually. It was a spectacular sight, and within a very few moments there was nothing to be seen below us. Well, almost nothing. One row of six or eight lights remained, apparently reflected in a canal running beside them. We waited and waited, but they stayed on, the only lights visible anywhere. The pilot flew us back and forth over them several times. After a while we could dimly make out the Parliament buildings very close to the lights. But why the canal? No canal existed in that part of Ottawa, running in that direction. I finally decided that they must have just washed the street on the Parliament grounds, and the reflection made it look like a canal. Apparently somebody at the Parliament buildings had forgotten a switch. It was a glorious example of accidental deceptive camouflage.

I recounted this experience to my air-force colleagues. It confirmed their suspicions that the government didn't even know enough to turn off its own lights.

On Sunday afternoon, 7 December 1941, I was in the Meteorological Office and our little radio was babbling on. I heard the name Hickham Field, which meant nothing to me, and something about an air raid. Then a reference to a place called Pearl Harbor, a name that I had never heard. I started listening carefully. It seemed as though something enormously important had happened. However, we received so many false reports during the war that we had learned to be sceptical about newscasts. But as the evening progressed and the facts came in, it was obvious that a momentous event really *had* taken place and that the United States and Japan were virtually at war. The next day it was officially declared.

Then Hitler made his greatest mistake of all – he declared war on the United States.

If he had used his brains he would have declared his neutrality in the Pacific war. Roosevelt would have had a terrible time with three groups of Americans who would have opposed America's declaring war on Germany. The first were the Bundists, Americans of German descent, and ardent pro-Nazis. The second were not German sympathizers but arch-isolationists – people like Colonel McCormick, the publisher of the Chicago *Tribune*, and Senator Wheeler of Montana. They and millions of Americans felt that all wars were bad and both sides in a war were equally wrong. They maintained that America should stay out of all foreign wars. The third group were military planners who felt the country was neither ready nor able to take on a war on two sides of the globe. America should concentrate her efforts on Japan and let Britain and the Commonwealth continue to go it alone in Europe.

When Hitler declared war on the United States he settled the matter once and for all. Roosevelt's opposition collapsed overnight.

We were thrilled. We had never doubted that we would win the war, but now it would be easier. Now we would also win any arguments with those Americans who had long maintained that there was no right or wrong in the two world wars. We were delighted to see vindicated those valiant and far-sighted Americans who had supported the cause of Britain and the Commonwealth, many of whom had come north to join the Canadian forces. Our sense of relief at gaining a great ally was tempered by the loss of lives, ships, and aircraft at the place whose name we now knew only too well.

Within a few days the British battleship *Prince of Wales* and the battle cruiser *Repulse* were sunk in the Pacific when a cloud layer cleared unexpectedly, exposing them to Japanese air attack. This stirred me to apply again to the navy. I was delighted to find that they now wanted a few met. officers. In February 1942, I was sworn in ("attested," in navalese) and started training in the local division, which was an old schoolhouse called HMCS *Carlton*. I trained three nights a week, generally getting the feel of the navy. I continued my duties at Rockcliffe while the traffic jam of new entries ahead of me gradually cleared. It was a constant source of exasperation to hear the news of the terrible happenings in Europe and to see Mackenzie King's government move so slowly when the freedom of the world was at stake.

Two months later came a great occasion. All my life I had had two ambitions for the next war. One was to take part in glamorous parades and the other was to win the Victoria Cross. I never won the VC nor any other decoration, but at least I got into a parade, although not one of great military significance.

It took place on a lovely warm spring evening, the kind when Canadians, having survived another winter, loved to sit out on their verandas and watch the world go by. Our whole division of new sailors marched through the residential area from our "ship" to the National Research Council. The lowly purpose was to see a film on venereal disease. I was given the exhalted task of being the officer who walks alone behind everyone else bringing up the rear. I felt that the eyes of the world were upon me and marched as smartly as I could, looking neither to the left nor to the right. All the small boys in the district walked along imitating me while the local dogs barked at my ankles. In due course we saw a most enlightening film and were dismissed to go home ignominiously in streetcars. That was the only parade I was ever in, either during or after the war.

In May I was called to active service and left my position at Rockcliffe. It meant ten days of full-time training at our school-house "ship." Much of the time was spent out in the playground tying knots. Apparently we were going to strangle the Germans. Those were the last knots I ever had to tie in the navy.

Another part of our indoctrination was reading an ancient volume called *King's Rules and Admiralty Instructions*. From it I learned two rules: naval officers were forbidden to carry brown paper parcels in public, and if a funeral passed by we were to salute the corpse.

Having thus been devirginized we were sent to Halifax for a month's further training in matters naval.

I was in a group of about twenty-five to thirty specialist officers. In addition to the gold braid on our arms, we had a narrow stripe of green to show we were Special Branch. We were all allegedly expert in one field or another.

One of my cabin-mates ("roommates" to landlubbers) was a most unworldly professor of classics from an American Ivy League university. He knew more Greek than Euripides, but we were never sure that he knew the difference between a motorcycle and a motor boat. At the conclusion of the course he was appointed senior intelligence officer at an important base in the Caribbean. I assumed he would be shot for incompetence within twenty-four hours, but he wasn't. Some months later I met him and asked how things were going. "Oh, fine, thanks," he said. "Tracking U-boats is the same sort of work I've been doing all my career." He was the type of brilliant scholar beloved by British Intelligence in two world wars, and we had inherited the tradition. Apparently the fitting together of incomplete scraps of information to form a pattern requires the same thought process, whether the material is U-boat transmissions or pieces of vases.

Another cabin-mate, an electronics expert, was a grizzled World War I army veteran. He told us a good deal about life in the trenches and gave us an intensive course in the literature and music of the period, all of it hilarious and all of it X-rated.

During our month at HMCS *Cornwallis*, the training base at the dockyard in Halifax, we were given a general introduction of varying degrees of merit. At 0600, before breakfast, we had an hour of training in signals. Our instructor was an elderly permanent-force sailor who, during the course of his career, had made the dizzying climb from Ordinary Seaman to Able Seaman, the next step up the ladder.

"I'll never get any higher," he lamented, looking at us with his watery bloodshot eyes. "I've done cells." He never told us what grievous sin he had committed to warrant naval prison.

Our chief occupation under his tutelage was memorizing the flags for a signal called the Boats' Recall. Our instructor stressed the importance of knowing this. Day after day we worked on the Boats' Recall. Finally, towards the end of the course, one of my landlubber Special Branch colleagues asked, "But how can boats see these flags if they're over the horizon?"

Our teacher replied wearily, "Boats don't go that far, Sir."

"What do you mean? They go all over the ocean."

"Not boats, Sir. Them's ships."

It finally dawned on our classmate what the rest of us had realized earlier. We were being taught the drill for battleships' signalling their lifeboats, very useful in the 1920s.

The mornings were spent bashing the parade ground. Evidently we were going to walk over the ocean the way the army was training to trample Germany underfoot.

One of our specialists had great trouble swinging his arms. This was an important part of smart marching. Of course, the right leg and the left arm must be swung forward simultaneously. This is a natural movement to most of the human race, and the navy, like the other services, just smartens it up a little. But that particular classmate, who knew everything there was to know about bacteriology, could never remember which arm went with which leg. He would start off correctly, but within a few steps would be swinging his right arm forward at the same time as his right leg, while his left arm worked in conjunction with his left leg. Thus he walked like a pacer (try it sometime). I can still see Petty Officer Barker taking off his hat and shielding his eyes as we marched past on our way to defeat the might of Germany.

Another feature of our training, which fortunately lasted for only one week, took place in a hell-hole called the Gun Battery. This was a long shed containing several large guns, to each of which about six or eight of us were assigned. We had to learn the different functions for every member of a gun crew, rotating in turn to do our different tasks. It was run by the devil himself, dressed in the uniform of a Chief Gunner's Mate. He screamed all afternoon while we bumped into each other and dropped dummy shells on our feet.

One day we actually got to sea. We went out with the mine-sweeping flotilla that had been working diligently outside the harbour since the start of the war without ever having found a mine (though they did later, as will be seen). At last I felt I was in the navy and loved every moment of it, except when somebody fired a test burst on a machine gun without warning right over my head as I was gazing out to sea. I jumped out of my skin. I had always hated loud bangs from when I was a little boy and would be the farthest away from any firecracker about to be set off. On passenger ships I was afraid to go anywhere near the whistle for fear it would blow. Why I joined the navy, the noisiest service, I have no idea, unless it was that like all Prairie boys, I loved sea stories. This was the only time I ever heard a gun of any size fired near me during the entire war.

All this time nobody showed us how to fire a depth charge nor told us a U-boat's cruising range, nor the depth from which they fired torpedoes, nor how we tracked them with radio direction finders. All in all, we received very little of value in a twentieth-century war. But if we ever became the signals officer in a British battleship that had lowered her boats for a harbour regatta, we knew how to get them back.

CHAPTER SEVENTEEN

Flashback: Signals from the Sea

Having been converted into an old salt in one month, I was sent back to Naval Service Headquarters in Ottawa. The navy, having recruited four met. officers, found it didn't know what to do with us. I was attached to the Staff Officer (Navigation) and performed a few odd jobs. In the office next door was the Staff Officer (Fuel). He was in charge of all petroleum products for the Canadian navy and felt he was long overdue for leave. When he found me sitting around doing nothing in particular he wangled permission to have me relieve him so that he could go on leave for two weeks. I knew nothing whatsoever about fuel beyond which gas station to stop at. However, he maintained that there was nothing to it and he would teach it all to me in a couple of days. Just before he went on leave I spent some time with him, and he showed me how to add certain figures to certain other figures to show the total of fuel in our eastern ports, and how to plot the convoy that was bringing oil from the Dutch West Indies. First thing Monday morning, as I sat at my new desk, a hurried message came from the Admiral's flag lieutenant. That very morning a conference had begun along the corridor, attended by senior officers of the British, Canadian, and American navies. It concerned the desperate situation on the Atlantic seaboard where the new U-boat attack was devastating the fuel supplies. "Would the Staff Officer (Fuel) please bring immediately statistics showing how the fuel consumption at the Canadian eastern bases for the last month compares with the estimates made six months earlier?" The admirals were waiting.

I was horrified. I hadn't the slightest notion where to find the information. I could have killed the regular Staff Officer (Fuel) for having put me in this position and I felt that I was about to lose the war singlehanded.

However, two days later I had rummaged through enough piles and done enough arithmetic to be able to send their lordships a reply, which seemed to quiet them down. Fortunately, the Staff Officer (Fuel) came back after his leave before I had done too much damage, so we eventually won the war.

I was then moved into the Operations Room, which was an intensely interesting place. This was where the huge wall maps of the North Atlantic showed the positions of our convoys and our guesses as to where U-boats were operating.

Soon after I joined the staff of the Operations Room I heard a story that I was unable to confirm. The Straits of Belle Isle lie between Newfoundland and Labrador and are notorious for fog. In peacetime ships used the route regularly as it was the shortest between Montreal and the UK. Some early convoys were sent this way until, on one terrible occasion, two met head on in dense fog. Thirty or forty ships were damaged in collisions before the convoys untangled themselves. The responsible operations staffs had failed to notice the impending danger. Whether or not this story is true, it kept us constantly aware of our convoys' projected positions, shown by string between pins on the chart.

After doing some more odd jobs, I was given the eminent post of Staff Officer (Coastal Convoys). I didn't make any decisions, but merely kept track of what ships were sailing from where to where, what ships were escorting them, and their position, courses and speeds, and also, unfortunately, losses. The U-boats were operating within sight of land in the lower St Lawrence River and gulf. In the summer of 1942 they torpedoed twenty-three ships in addition to others sunk off the East Coast. This was not only a serious new development but also a considerable embarrassment to the navy and government. A secret session of Parliament was called to debate the subject, and this created a great flap in the Operations Division as graphs, charts, and alibis were prepared. The reasons for the German successes were, first, the daring and skill of the U-boat commanders; second, the widespread nature of our responsibilities and the shortage of escort vessels – the hard pressed Allies were unavoidably spread too thin; and third, the fact that many freshwater rivers flow into the St Lawrence, setting up planes of varying degrees of salinity, driving our asdic (sonar) operators to distraction.

Each day we pinned on the wall chart little red submarines representing the approximate locations of U-boats, as estimated by the very bright lads in the Admiralty tracking room in London. Their educated guesses were based on a variety of information, from air photos of shipyards in Germany to radio intercepts from the U-boats themselves, which were too chatty for their own good.

Thirty-five years later, Eric McLean, Canada's most distinguished music critic, told me a story in this connection. At one point during the war, he was a very young and apparently very green radio operator assigned to detect U-boat transmissions from a remote post in Newfoundland. One night during his lonely vigil he was startled to pick up a voice addressing him in German-accented English. The submarine was obviously close to shore. Its captain told him they had changed procedures and that Eric was listening on the wrong wavelength. After giving him the correct frequency and wishing him a cordial goodnight the voice vanished into the darkness while the red-faced future Order of Canada recipient retuned his set.

In the fall we moved into a new building specially built for the war. We had been in an old apartment block until that time. The Operations Room was bigger, and now, instead of pushing pins into beaverboard, we used little magnets that stuck to the metal backing of the wall charts.

At the side of the main plot in our new operations room was a little ledge that could hold two numbered cards giving the total of enemy submarines in the North Atlantic that day. At first, two digits were quite sufficient. But the total kept increasing despite an occasional sinking by our hungry antisubmarine ships. One gloomy day we had to call in the signmaker and have him enlarge the ledge so that three digits could be accommodated. We used all three places from then on.

The plot looked quite pretty, with its cute little toy submarines clinging to the walls by their magnetic bottoms. It was hard to remember that each one was a deadly enemy, and that the total force was in imminent danger of strangling the only free country in Europe, our only possible base for bombers and invasion forces.

I was made a regular watchkeeping officer, which meant shift work. We kept track of all the ships in convoys in the North Atlantic. We lived a pleasant life, but our hearts were always at sea. Signals were constantly coming in, typed on pink telegram-size paper, written in navalese and kept as short as possible to make it harder for the enemy to home in and locate the source of transmission. The convoy torpedoings were at their worst in mid-Atlantic and it was a grim task to take ship after ship off one part of the wall plot, where the convoys were listed, and put them in the "ships sunk" column. We would receive such cryptic messages as "SEABELLE SUNK 6 U-BOATS IN CONTACT AM ATTACKING," followed by the weather group. This usually reported incredibly high winds as the North Atlantic waged its own war on us.

Once a convoy was a hundred miles or so south of Iceland and we received a report of a sinking that concluded with the following message: "RESCUED SEAMAN WALLENHUPT SURVIVOR FROM SS PUERTO RICAN CONDITION GOOD." It was unusual to have an individual sailor's rescue reported in an operational signal. We couldn't quite see the reason until we started investigating. The *Puerto Rican* was not part of that convoy. However, we had vague recollections of having seen her name and hunted back to find when the last convoy sailed through that area. We finally discovered that she had been sunk two weeks earlier. Two years later an account appeared in *Reader's Digest** describing how Seaman Wallenhupt survived in an open boat south of Iceland in mid-winter.

One of the worst battles took place just before the end of 1942. I was on duty throughout the night of 27–28 December, as sinking after sinking came in. I went home in daylight to bed and the next day was married to my college sweetheart in Toronto. When I returned from our honeymoon I found that the convoy had been one of the most severely mauled of the war.

One night I received a signal on white paper with a red border. This meant more trouble. It was headed "IMMEDIATE AND MOST SECRET." It briefly reported the sinking of the Canadian corvette *Weyburn* in the North African landings. I reported this to my senior officer who asked me to inform Rear Admiral Jones, Vice Chief of the Naval Staff, who was directly responsible for operations. We had to be very careful about security in telephone messages, not just from fear of helping the enemy but, in a case

---

* "Captain of His Fate," *Reader's Digest*, July 1944, 45. I am indebted to Miss Laura Giangrande of *Reader's Digest* for locating this forty-five-year-old article.

such as this, to avoid leaking the information to the press before the next of kin could be informed. Hence my conversation with the sleepy Admiral Jones at about two A.M. went as follows:

"Sir, this is Sub-Lieutenant McElheran, Duty Operations Officer. We just received a signal. You know the way she burns?"

"What?"

"The way she burns, in Saskatchewan?" (referring to the western town after which the corvette was named.)

"Oh, yes, yes. What about her?"

"She took a dive."

"Oh. I see. (Pause.) You mean, bumped off?"

"Yes, Sir."

"Oh. Thank you very much. Good night."

"Good night, Sir."

I doubt whether this absurd conversation about a grim subject in an improvised code would have fooled any German cryptographer.

One morning I had just returned to my boarding house and had settled down for a day's sleep when the landlady told me I was wanted on the phone. My immediate superior, Lieutenant Commander Atwood, was on the line and did not sound completely charming.

"McElheran, come down here right away."

"Yes, Sir." This sounded bad, and in fact it was.

As stated earlier one of our duties was to place little red submarines on the chart at the positions estimated by our intelligence officers. Our operations superiors kept an eye on the plot to make sure that we weren't sending any convoys into areas where we suspected there were U-boats. I hurried into the Operations Room, heart in mouth. A grim-looking Lieutenant Commander Atwood said, "Take a look at this U-boat position. Did you put it here?"

"Yes, Sir." I replied, noting a little red marker just south of Newfoundland. It was very close to a convoy consisting of just one ship and one of our smallest escort vessels.

"Check that position again," he said, curtly handing me a signal. I did so and found to my horror that I was a full five degrees of longitude, or 250 miles, too far west. I had been so intent on getting it accurate to the nearest minute that I had mistaken a blurry-looking three for an eight. The Staff Officer (Operations) had sent the small convoy a signal in the night warning them that a submarine was in their immediate vicinity. I was appalled, but was at least relieved that it hadn't been the other way around, a mistake that could have proved disastrous. After a blast about being careful, I was permitted to go back to bed.

Two or three years later, I met an officer who had served in the one and only naval vessel that had escorted the one and only ship in that small convoy.

"Do you remember one night, when you were south of New-foundland, receiving a signal that there was a U-boat in your immediate vicinity, and then a few hours later another saying, 'Cancel my last signal'?"

"Boy, do I! What a flap that caused."

"Sorry about that."

Another incident, this with tragic consequences, took place in that same area. Again, it concerned a convoy of one ship escorted by a single Canadian antisubmarine minesweeper. The night was dark and foggy. All eyes on board were straining to pierce the murky void. Suddenly a surfaced submarine was seen a short distance ahead. The minesweeper flashed the recognition signal and increased speed. There was no reply, and in seconds the Canadian ship rammed the submarine, which sank almost imme-diately. There was much rejoicing aboard the little escort vessel, especially when a body floated to the surface. U-boats were known to be very good at feigning having been sunk, often releasing oil, wreckage, and even animal flesh to throw off pursuit. Thus the sinking of a U-boat was only considered certain by the Admiralty if the crew were seen to abandon ship or human remains were found. As the triumphant minesweeper's crew hoisted the body aboard as proof of their victory, they were appalled to note that the corpse was wearing the type of overall issued to crews of British submarines: this had, in fact, been a British sub in transit to Halifax. It was being loaned to the Canadian navy for training purposes, escorted as a safety pre-caution by one of our ships. They had become separated in the fog. Why it did not flash the recognition signal will never be known, but one opinion was that the signalman may have been in such a hurry that he banged the lamp against something and broke it.

Investigation disclosed that all appropriate authorities had been warned about the submarine's trip, with one exception – the minesweeper. Someone had not noticed the little convoy on the plot.

A grisly story in this connection concerns a Canadian corvette that sank a U-boat early in a transatlantic crossing. They collected fragments of flesh that they were sure were human. However, the Admiralty, notoriously hard to convince, would doubtless insist on testing the gruesome remains before crediting the corvette with the sinking. The crew unanimously voted to clear out the ship's refrigerator and put the pieces inside to preserve them for the Admiralty's edification. This, of course, meant eating canned food and ship's biscuits for the rest of the voyage, but the little ship eventually received credit for her kill.

Not all our signals, however, were depressing. Every book on naval warfare has its favourite set of humorous signals, usually deliberately witty. My all-time favourite was sent in deadly earnest.

Canadian shipyards were building a series of antisubmarine vessels. These were named after islands in the western hemisphere – *Anticosti*, *Porcher*, *Prospect*, etc. HMCS *Prospect* was proceeding down the St Lawrence from the builders to join the navy when she damaged her propeller or screw in the locks at Cornwall, Ontario. Divers managed to remove the ruined propeller and straighten the shaft, whereupon we received the following signal:

TO: NAVAL SERVICE HEADQUARTERS OTTAWA
FROM: SENIOR NAVAL OFFICER CORNWALL
IMPORTANT
PROSPECT READY FOR SCREW REQUEST INSTRUCTIONS

This signal soon had an unusually wide distribution.

Our happiest day, however, was in May 1943. A convoy was taking a beating in the central part of the ocean. This was the favourite hunting ground of the U-boats because they were out of range of shore-based aircraft. Thus they could operate on the surface away from the convoy and manoeuvre into position without fear of air attack. On this particular day we received a garbled transmission indicating that an aircraft actually was present and had sunk a U-boat. We checked the position again and it didn't seem possible that anything could be flying in that area. However, one or two more signals, including a few from ship to aircraft, indicated that this was indeed so. We soon found out that we had obtained some of the great American planes called Liberators. They were flying from Iceland and Newfoundland and could now cover that terrible gap.

Postwar totals indicate that in March and April, twenty-seven U-boats were sunk in the Atlantic, over half by aircraft. The dramatic drop in Allied shipping sunk during the spring of 1943 was as follows:

Month	Tonnage Sunk
March	514,744
April	241,687
May	199,409
June	21,759

At that point we won the war.

CHAPTER EIGHTEEN

Flashback: Mines, Queens, and Fogs

After I had served in Ottawa for a year, the navy decided that it needed met. officers after all and sent me to Halifax. I now had a beautiful new title: Staff Officer (Meteorology) on the staff of the commander in chief Canadian North West Atlantic. It was June 1943. My lovely bride of six months and I packed up and headed east.

The day I reported to the Operations Room for duty, 2 June, things were tense. The great port of Halifax, a vital link in the supply chain that sustained the Allied war effort in Britain, was closed due to enemy action. For the first time in two wars, German mines had been found the previous day in the approaches to the harbour.

In writing of this important but little-known event in Canadian history, I had great difficulty in finding any references to it, let alone details. However, in October 1994, I received a package of fascinating documents. Mr D.G. Gage of the Friends of the Canadian War Museum had finally located a number of reports. It appears that the documents had first been classified as "secret" to avoid giving information to the enemy and then, in the well-known manner of Important Documents, had hidden themselves deep in the files only to reappear when least expected.

A German submarine had laid a large number of mines in the approaches to Halifax harbour during the night of 30 or 31 May 1943. These were moored magnetic mines that exploded when activated by the approach of a steel ship's magnetic field. Each was powerful enough to sink a large vessel. They contained time-delay mechanisms so that it was necessary to sweep over a mine a number of times before it exploded. However, conventional minesweeping gear could cut the mooring cable and bring it to the surface first try, where it could be sunk by gunfire. Each one had booby-trap devices, designed to blow up if anyone tried to explore its insides. Fortunately this nasty welcoming device didn't always work.

One maverick mine showed another peculiarity. Rising to the surface after its cable was cut, it floated high in the water, while the nearby minesweeper's crew admired their handiwork and prepared for the execution. The mine, defiant to the end, suddenly poured out streams of smoke, then gracefully sank beneath the waves, to the astonishment of all.

On 1 June a convoy of ten ships sailed for Boston. Three escort vessels steamed ahead on routine antisubmarine patrol. Suddenly they sighted three floating mines. They routed the ships safely around them and stood guard, signalling the Operations Room in Halifax. The convoy sailed on, blissfully ignorant of the fact that they were passing over Canada's first enemy minefield. The magnetic mines had apparently not yet been sufficiently stimulated. Faulty depth controls had caused the three strays to rise to the surface.

Three of our own defensive mines were known to have broken loose recently, and it was felt that these had at last been found. However, even though no enemy mines had ever been discovered in Canadian waters, the Staff Officer (Minesweeping), Lieutenant Commander Barkhouse RCNR, was suspicious. He contacted Lt George Rundle, (who bore the unwieldy title "Render Mines Safe Officer") and they took a fast launch out to inspect the mines.

At this point the official account is vague, saying only that they ascertained that the mines were German. The story that was common talk in the Operations Room was that Barkhouse and Rundle climbed into a small boat and rowed to one of the mines. While the former steadied the dory in the tossing sea, Rundle delicately unscrewed the cover over the mechanism and reached in to deactivate the mine. As he did so, the firing pin snapped down on the back of his hand. If it had gone through his fingers and hit the firing cap the two men and probably the nearby minesweeper would have been blown to pieces. As it was, despite the choppy sea he managed to unscrew the pin and disarm the mine.

The famous port was officially closed at 9:18 P.M. on 1 June. Minesweeping operations in the channel commenced at dawn, 2 June. By evening a channel had been cleared and a convoy sailed safely to the open sea – a tremendous but little-heralded victory.

Sweeping continued in the days ahead. On two occasions mines were towed and beached by Fairmile motor launch 053, commanded by Sub-Lieutenant Schuthe RCNVR. The intrepid Lieutenant Rundle dismantled the firing device; luckily, the booby-trap mechanism was defective. In all, three mines were recovered intact. One was sent to London and one to Washington for further study.

While all this was going on in the foggy sea a few miles away, the operations staff organized a sweepstake to relieve the tension. I held ticket number 43. As the grim total of mines destroyed climbed day by day it appeared for awhile as though I might win my only raffle, but after a few hours the total resumed its steady climb well past my number.

Many ships were pressed into service, including several belonging to the British navy. At one time twenty-five ships were sweeping simultaneously. By 25 June, fifty-five mines had been destroyed, thirty-seven by the gallant little wooden-hulled magnetic minesweepers. The only ship to be sunk by the mines was a 2,000-ton American freighter, SS *Halma*, which took a forbidden short-cut across an unswept area and went down within sight of land.

I have not found any explanation as to how a U-boat could wander around all night on our front porch without being detected.

Rundle was awarded the rare George Medal for his gallantry, and Barkhouse and Schuthe were given the honour known as "Mentioned in Dispatches."

In the fall of 1943 the British Admiralty Minesweeping Summary No. 198 had this to say of the Halifax incident: "The whole operation was most successful, particularly when it is considered that the minesweeping forces were confronted so suddenly with such a large minesweeping problem, after a long period of minesweeping inactivity, and also taking into consideration that the operations were continually hampered by fog."

My new job was to act as liaison between the navy and the civilian meteorological office, run by my old outfit, the Department of Transport. Each day I was given a tracing of the weathermap and was briefed by the duty forecaster. I then took the map around the operations staff from the admiral on down, after which I toured the dockyard. I visited the escort vessels as they came in, corrected their aneroid barometers (which were usually badly out of whack from depth charges or gunfire), and had a chat with the navigation officer. I would ask for comments on our forecasts and encourage him to add coded weather reports to any messages sent from sea as we had no other means of knowing the weather conditions beyond our own visibility limits.

It was fascinating and gripping to deal with these men and ships during their brief respites from the Battle of the Atlantic. The escort vessels painted in the zigzag patterns of white, pale blue, darker blue, and green, and their sirens, sounding a little like yelps, were constantly heard as they manoeuvred in the harbour. Upstream past the dockyard was the famous Bedford Basin, a huge area several miles long, capable of holding all the world's navies. It was always well dotted with large gray merchant ships waiting to form the next convoy. Every few days a stately procession lasting several hours would pass in or out of the harbour.

Whenever a convoy was due to sail, I would brief the officer in charge with the latest forecast. He would take the weathermap into a convoy conference attended by all the grizzled merchant captains. They spoke several languages and sailed under several flags, but were all united in their determination to fight their way through the U-boats and deliver their cargoes, so vital to the cause of freedom.

It was sometimes hard to assess how much help this information was to the mariners, but one night we had positive proof that we *were* earning our keep.

One of the officers whom I briefed every morning was the captain of the dockyard. During the night a merchant ship carrying an enormously valuable cargo caught fire. The captain of the dockyard turned out and took command of the operation. When the fire was extinguished, it was obvious that the freighter had to be beached and her cargo removed. The old salts who ran the salvage tugs wanted to tow her to a nearby bay, but the captain of the dockyard said, "No. There's going to be a gale coming in on that side. We'll lose the cargo and the ship'll break up." The old salts sniffed the wind and looked at the moon and said, "There isn't going to be any wind from that direction."

"Well, my met. officer says there *is*, and therefore we're going to take her around and put her over there where she'll be more sheltered."

He pulled rank and, despite the grumblings of the local experts, insisted on their towing the ship a considerable distance to another bay. When I saw him a day or two later he thanked me. He said that the unloading arrangements were proceeding satisfactorily, they had salvaged the cargo, and that if she had been beached in the first location ship and cargo alike would have been lost.

I did a rough estimate and calculated that the value of the salvaged cargo plus that of the ship was at least equal to the expenses for one year of the entire Meteorological Service of Canada.

During the summer of 1943, we were frequently visited by one or other of the two largest ships afloat, the great Cunarders *Queen Elizabeth* and *Queen Mary*, both of the Cunard line. They had been converted into troopships, their luxury fittings removed, their swimming pools drained, and every available square foot covered with triple-decker bunks. They carried eighteen thousand men and two thousand crew. They were so fast that no antisubmarine vessel could keep up with them, so they sailed across the Atlantic unescorted, zigzagging at irregular intervals to make their movements harder to anticipate. Neither was ever touched by the enemy. In port their massive grey bulk dominated the landscape.

The twenty thousand human beings were not the only living creatures aboard the great Queens. After one visit we found that there was no rat poison for sale anywhere in Halifax: the ship's supply officer had bought every case in the entire area.

In the 1930s our family heard a story about the naming of these ships from a highly placed source who cannot be revealed. Up to that time all Cunarders had names ending in "ia" – *Lusitania*, *Mauritania*, etc. When the company was selecting a name for their newest and greatest ship they hit upon the bright idea of calling her *Queen Victoria*. This, however, required royal permission. The head of the company visited King George V.

"We would like your Majesty's permission to name our greatest ship after England's greatest queen."

His Majesty graciously replied, "I'll speak to my wife about it, but I am sure she will be delighted to have your splendid new ship named after her." Thus the company had no alternative but to call the new liner *Queen Mary*.

Sometime later, a newer and slightly larger sister ship was built. King George V had passed away, King Edward VIII had come and gone and George VI was now on the throne. The company felt it diplomatic to name the ship after his consort – hence *Queen Elizabeth*. There is still no *Queen Victoria*.

Manoeuvring these giants in and out of the port was a tricky business. The entrance to the Halifax harbour is narrow and in wartime was made even smaller by antisubmarine nets extending from the shore on each side to two small ships permanently anchored fifty yards or so apart in the middle of the channel. A cable suspending the nets was normally in position between the two vessels, completely sealing off the harbour from submarines. The gate was opened when the proper identification signals were received.

Sailing a mammoth ship through this narrow gap without hitting one of the gate vessels required expert seamanship and good visibility. The jetty at which all large ships tied up was behind the Nova Scotian Hotel. This was just across a little park from our headquarters building, which included the civilian met. office and the Operations Room. Between the hotel and the water was a narrow strip of land on which was situated the railway station. When a Queen was in port, there was a constant succession of long railway trains coming in beside the ship, disgorging hundreds of American soldiers, and then backing out to make way for another train. The loading process took two days. At sailing time there was much huffing and puffing and tooting as a gaggle of tugs pulled the Queen out and turned her around, aiming her at the gate. She then steamed out cautiously by herself.

One of the problems with being a weatherman in the Halifax area is fog. The cold Labrador current flows south along the coast. When air masses that originate in the general neighbourhood of the Gulf of Mexico drift across the chilly water, they are cooled below their capacity to hold moisture in transparent state, and sea fog forms. In much of the spring and summer there is dense fog off the coast extending all the way to Newfoundland and beyond. At night this frequently drifts inland and is usually blown out to sea the next morning or warmed sufficiently by the sun to evaporate. If an arctic air mass decided to move down from the northwest, it was relatively easy to predict its arrival. It would push the fog out to sea and we would have a few days of clear dry air, blue skies, and unlimited visibility.

If no polar air mass showed up it was extremely difficult to predict whether or not the harbour would be fogged in. There was no such thing as a compromise – it was visibility either of several miles or less than a hundred yards.

On one occasion the fog moved in when one of the queens was loading. On the morning she was due to sail the forecast called for early fog and then clearing. Mid-morning came and went, but the fog persisted. The captain delayed the sailing, not wanting to risk going through the gate and swept channel under such conditions.

Messages flew back and forth. I spent a tense and busy day running forecasts from the Department of Transport Meteorological Office to the Operations Room, whence they were relayed to the ship. There was every indication that the sun would do its good work and burn off the fog as the day progressed. Light wisps of smoke could be seen floating up from the pale gray funnels as she raised steam for her exit. Visibility wasn't too bad, but in the captain's mind it was not good enough. There was no point in running risks when the forecast said it was going to clear at any moment.

"It can't last much longer," was our official prognostication. But by mid-afternoon, contrary to all the laws of physics, it had thickened. It was now impossible to see the ship from our building. As late afternoon came on, the visibility was terrible and the Queen settled down for the night. The met. office felt sure it would clear the next morning, and so the message went out.

But next day the fog was even denser. Again questions from the ship, and again answers based largely on faith in the Sun God. But he showed no sign of concern or even awareness of our desperate problem. The white pall persisted throughout the day. Meanwhile, twenty thousand hungry men and, presumably, a large contingent of rats were working on the food supplies that were supposed to last for a four- or five-day voyage. We studied the map, looking for a cool dry air mass heading our way, but no such friend was anywhere to be seen. It was a typical stagnant situation that in the central part of the continent means a humid heat wave, but in Halifax a clammy damp fog. The third day dawned, and the fog was thicker than ever. We were frantic and so was everybody else. Finally, a stentorian blast indicated that the captain was fed up and was going home anyway, fog or no fog. Higher-pitched toots from the tugs told us that they were in action. Looking across from our building we could see the top parts of the Queen's masts and funnels starting to move. They extended above the densest layer of the fog. There was no trace of her hull or superstructure. We could tell that she was being turned around and aimed towards the harbour entrance. She finally groped her way out on radar, the low C of her awesome fog horn making the whole city tremble. She disappeared from sight and went on her way, not only through the narrow gap in the antisubmarine net but through our swept channel out to open sea. She stayed clear of our own minefields by picking up the channel marker buoys on her radar, a very new device in those days. It was phenomenal seamanship. We were all nervous wrecks until we received word that she had cleared the minefield and was safely in the open ocean.

One little-known story illustrates the prevalence of fog on the Grand Banks off Canada's eastern coast. A certain Newfoundland lighthouse warned ships away from its rocky promontory with a fog gun instead of a horn. This was mounted just above the light-house keeper's bedroom. It fired automatically once a minute, shaking the whole lighthouse. On one occasion, after three days and three nights of dense fog, the mechanism failed. As the sixty-second mark went by in silence, the keeper, who had been sleep-ing soundly, woke with a tremendous start, looked wildly around the room, and shouted, "Great Scott! What was that?"

Like most Newfie stories, probably the only true part is the fog.

Early in August 1943, I was aware that an approaching Queen in mid-ocean was receiving an exceptional amount of attention from senior officers, who visited the plotting room much more frequently than usual. They also would question me about the weather in her vicinity and were particularly interested in whether or not there were any strong head winds. This fidgety behaviour aroused my curiosity. Obviously, *somebody* was aboard.

When, on 9 August, the *Queen Mary* eventually arrived, the secret was out. Every tugboat crew member spotted on the bridge a familiar figure wearing a reefer jacket and waving a cigar at the whole world. Winston had arrived! He was on his way to Quebec City to confer with Roosevelt and others on the future conduct of the war, especially the plans for the invasion of Europe.

Despite our top security, everybody in town knew as soon as the tugboat crews landed. In his memoirs, Churchill says he was greeted by a large crowd and led them in singing "The Maple Leaf Forever" and "O Canada." We never heard that story until it appeared in his memoirs. Unfortunately we missed what was presumably Churchill's entire career as a choral conductor.

Within a few days we received word from the Americans that a tropical hurricane was moving up the Atlantic coast. It was centred off Boston, where it gave a rough time to the ship that had signalled its last position. We had been watching this storm for some days and knew it was heading straight for us. Since the start of the war it had been forbidden to publicize our forecasts in the Maritime provinces. We knew that U-boats could gain much useful information if they could receive our civilian weather reports. However, it was permissible to break this rule on the approach of an exceptionally severe weather condition. Rube Hornstein, head of the civilian met. office, called me in one day and showed me the chart, saying, "McElheran, we've got to get permission to issue a public warning." I spoke to the commander in chief, Rear Admiral Murray, and he gave his consent for us to break the customary silence. Out went the warning and the news media jumped on it with great delight, as they had been starved for weather news for years. In the streetcar on the way home I could hear everyone buzzing with excitement about the hurricane scheduled to hit the next day. I felt very proud of being a part of such a flap. Next morning dawned, and it had the hush that one expects before a terrible storm. Not a ripple. I went into the met. office.

"Where's our storm?" I asked.

"Well, it must be there somewhere, because it was there yesterday," was the reply.

There were no other reports as no ships in that area had broken radio silence; nor did we have aircraft at that time to fly into the storm. Southern Nova Scotia was not reporting any trouble, but that merely strengthened our opinion that this storm was going to be a heller once it hit. People were still talking about it on the streetcar that evening.

"Where's this storm they're warning us about?"

"I don't know, but it'll be terrible when it hits us."

I was pretty uneasy now about my hurricane. I went to bed hoping to have the roof blown off before morning.

The next day was lovely, with just a gentle breeze. Roof still on. We issued a shamefaced cancellation. But what had happened? It had definitely been a bad storm and it had definitely been headed straight for us, and there was definitely no trace of it anywhere now.

A few days later, just as our nerves were beginning to settle down, the British battle cruiser *Renown* came in and tied up across from us. Within ten minutes an irate lieutenant commander RN stormed into our building and demanded to see the staff met. officer.

"What can I do for you?" I respectfully inquired, noting that his sleeve bore two more stripes than mine.

"I'm *Renown's* met. officer. What do you people mean, telling us that we're going to have light winds when we run into a hurricane in mid-ocean?"

I took him in to Rube Hornstein and we examined the maps for the days in question. The Englishman showed us the exact position where *Renown* had run into the storm. It had all the characteristics of a tropical hurricane, including the calm eye in the middle. The wind had been recorded up to 120 knots and then the anemometer (wind gauge) blew away. The ship had suffered much damage to her superstructure. We presented our alibis and apologies to our outraged colleague. All agreed that there was no recorded case of a tropical hurricane's moving from one confirmed position to another at so high a speed, nor had one ever been reported that far out in the North Atlantic. *Renown's* met. o. left in a more benign mood, impressed at having sailed through a record-breaking tropical hurricane.

The reason for *Renown's* visit was to take Churchill back following the Quebec conference. After she had sailed with her distinguished passenger, a story circulated to the effect that Roosevelt had come to see him off, and as the two leaders were driving past a pretty park on Spring Garden Road they decided to stop and go for a walk. A drunk lying on a bench sat up, took one look, and gave up drink. This account was widely believed – at any rate, the part about Churchill and Roosevelt walking through the park. What none of us realized until after the war was that Roosevelt was unable to walk. The media had done such a tasteful job of presenting the great president to the world that it was years before we learned that he had been a polio victim and lived in a wheelchair.

One day in October Rube Hornstein pointed to some circles on a map and said, "Brock, we've got another hurricane coming."

"Rube, do you really think this one's going to make it this time?" I asked nervously. I reminded him of the old mariners' maxim: June – too soon; July – stand by; August – come they must; September – remember; October – all over.

"I'm sure of it. It's headed right this way," he replied with confidence. I immediately went to Admiral Murray and said, "Sir, the met. office would like permission to issue a hurricane warning."

"What, again? Are you sure you want to?"

"Yes, Sir. They feel we must warn the fishing fleet and others who could suffer damage."

He eyed me sternly for a few moments and then said grimly, "All right. You have my permission, but I hope you're right this time."

Out went the forecast to the newspapers and radio, and once more they pounced on it with delight. Once more I heard people in the streetcar saying such things as, "Well, they're talking about another hurricane. I wonder if this one'll bother to show up."

The next day was the usual nerve-wracking period of calm, checking our charts, sniffing the wind, and looking for every little sign that might indicate that we were, after all, correct.

The third day, to our enormous relief, we had a good blow, although technically not quite of hurricane force. Nevertheless it kicked up a fuss and even brought down a few tree branches, much to our delight. Weathermen have a peculiar attitude towards storms.

Next morning Rube greeted me with "What about 'October, all over'?"

No answer.

CHAPTER NINETEEN

Flashback: Pitch, Roll, Pitch

Since May 1943, when the Liberators had appeared over the mid-ocean gap, the U-boats had been quiet in the North Atlantic. In September, I convinced my bosses that the war could not continue to operate satisfactorily unless I visited the base in St John's, Newfoundland, which was in our command. They agreed, and I found a ride in a RCAF transport plane. We flew to Gander on an overcast day. As we let down through the cloud layer a steep hill rose directly in front of us into the clouds. We veered sharply to starboard in the nick of time. Later that day I was visiting the met. office when one of the flying-control officers said, "You're lucky to be alive. Your pilot brought you in on the wrong leg of the approach. We thought you were going to crash into that mountainside. We had no way of warning you." (Aircraft radios were scarce in those days.)

I visited the big naval base at St John's. This was where the mid-ocean escort vessels turned around after handing over their convoys to local forces for continuation of the voyage to Halifax, Boston, or New York. For a better understanding of life at sea I had arranged to return to Halifax in the senior ship of one of these escort groups. I stayed in an officers' dormitory the night before we were to sail. One of my roommates, asking when I was leaving, said, "That convoy's been badly beaten up. The Jerries have attacked for the first time with acoustic torpedoes, which home in on the sound of the propellers and thus can't miss. Several escorts have been sunk, and they've combined with another convoy to enable them to use both escort groups. It's really quite a nasty situation."

Up until now, escort vessels in the North Atlantic had been virtually untouched by the enemy, chiefly because they were too fast, too manoeuvrable, and too shallow to be easy targets. Also, they were of much less value than the ten-thousand-ton freighters and tankers that were taking the materials of war and life to Great Britain. This was an unpleasant new development.

I sailed in HMS *Leamington*, the senior ship of a small group of escort vessels. She was one of the fifty World War I destroyers the Americans traded to Britain for a ninety-nine-year lease on several British bases. They were all four-stackers, very narrow, and notorious for rolling badly in a rough sea. *Leamington* had been part of the escort of the ill-fated convoy PQ 17, which was devastated while attempting to take supplies to the Russians via the northern route.

As we sailed through the narrow mouth of the historic harbour of St John's we entered a world of dense fog. Our fellow escorts astern dissolved into white. We could feel the swells of the ocean causing us to pitch, a lovely gentle motion that I much enjoyed. After a while I found it a trifle hard to focus my eyes and soon decided I was about to die. In due course up came my lunch. This experience greatly increased my admiration for those landlubbers who went to sea for the duration of the war and carried out their arduous duties in seas much worse than this.

All night and the next morning we steamed on as though alone in the world. Our captain, Lt A.D.B. Campbell, RN, talked to the crew over the loudspeaker system. He told them about the acoustic torpedoes and that we were heading into a new and dangerous situation. He was calm and frank – a very young Royal Naval lieutenant who was obviously highly regarded by his men.

A Canadian officer attached to the ship had bought some peanut butter in Newfoundland. At teatime he passed it around to the British officers. The captain spread some on a piece of bread and ate it musingly. He told us it was the first time in his life he had eaten this American invention. "Now I can write my wife and tell her I've eaten peanut butter." Then, after a thoughtful silence, he added with the wistfulness of a man who had been away from his loved one far too long, "Aren't wives wonderful! Who else would care whether or not you've tasted peanut butter."

Late on the second afternoon our calculations indicated that we should be near the convoy, and soon our radar picked up the forward ships through the fog. These were the two convoys that had been combined, ONS 18 and ON 202. We saw nothing except the blips on the radar screen. Our captain ordered his escort vessels to station themselves at different places around the convoy. We moved towards the senior ship of the escort, a smaller blip in front of the big merchant ships. She was a British destroyer, HMS *Keppel*. Her captain, Commander M.J. Evans, RN, was senior to ours and as we approached, our captain asked me to take off a turtleneck sweater that I was wearing under my jacket. He didn't want the senior ship to see one of his officers in this slight deviation from the dress code. So, off came the sweater and the shivers increased in the damp fog, but our honour was preserved.

We swung in alongside the larger destroyer and ran parallel about fifty yards apart while the two captains chatted through their loud hailers. The commander of the mid-ocean escort was obviously tired and had been through one of the worst battles of the war. Several merchant ships and escort vessels had been sunk by acoustic torpedoes, with heavy loss of life. In return, *Keppel* had rammed and sunk a U-boat. Nevertheless he was calm, suave, and thoroughly polite. So was our young captain. The senior man's last shout was to ask whether Newfoundland's time was still "minus three and a half" behind Greenwich mean time. *Keppel* then increased speed and disappeared into the fog.

There we were, far out to sea, in command of two large convoys of which we could see absolutely nothing. I found this whole demonstration of technology and seamanship miraculous – all the more so when I remembered the horrors that those crews had survived and that might appear again at any minute. We continued pitching into the sea, our bow crashing and hissing with every wave as the light faded and night settled in.

I slept on a couch in the wardroom. At about midnight the bells suddenly rang "Action Stations." In accordance with my orders, I ran up on the bridge. The crew were running around the darkened ship hurrying to their posts. Nothing could be seen in the blackness ahead. We accelerated to eighteen knots, which we later found was the speed most vulnerable to attack by the new torpedoes. Our radar had detected an echo ahead of us, and we were giving chase. It disappeared, as one would expect if a submarine submerged. In a minute or two we had reached the estimated position of the target and fired off a pattern of depth charges. In a few moments the destroyer clanged from stem to stern as the charges exploded underwater. It was as though we had been hit by a giant sledgehammer. Ghostly columns of white water rose slowly astern of us in the dim light. I was glad to be well away from the end of the ship that attracted acoustic torpedoes. The depth-charge crews at the stern were doubtless wondering whether they would be blown sky-high at any minute. We circled around but saw no wreckage or any other evidence of a U-boat. To this day we don't know whether we had actually attacked one or had merely chased a phenomenon that often fooled radar operators, a fog echo.

Next morning we sailed on, still in fog, still no sight of the convoy. But in a day or two, as we approached Halifax, suddenly the fog cleared and there, astern of us, was a glorious sight: fifty-six grey merchant ships in perfect order, rolling gently, steaming ahead relentlessly as though nothing had happened. Soon we entered Halifax harbour without further incident. I felt that I was now a little more a part of the real navy.

Postwar accounts report that out of the sixty-three merchant ships in our convoy, seven were torpedoed, and six of these sank. As for the escorts, on 20 September the British frigate *Lagan* had had her stern blown off but did not sink. A short time later another British frigate, *Polyanthus*, and the Canadian destroyer *St Croix* were sunk by torpedoes that hit their sterns. Eighty-one survivors of *St Croix* were rescued by the British frigate *Itchen* thirteen hours after their shipmates had gone to the bottom. The delay was due to the compulsory hunt for the U-boat. One of *Polyanthus*'s crew was rescued by *Itchen* the next day. Two days later another acoustic torpedo sank *Itchen*. There were only three survivors, one each from *Itchen*, *Polyanthus*, and *St Croix*, the latter two having of course been through this terrible experience twice.

The acoustic torpedo created quite a flap for a week or two, but the Admiralty, having anticipated this gadget, soon had a cure. Ships were given a pair of steel bars to be towed well astern of the vessel. They made a chattering noise which the acoustic torpedoes thought was the sound of propellers. If they hit the bars at all they did no damage to the ship. These were variously called "chatter bars," "foxer," and "CAT."

Winter came and lasted an eternity – a succession of snow storms, convoys, thaws, gales, clear polar days, escort vessels, queues for movies, sailors of many nations crowding the side-walks, barometers, friends passing through on their way to war, slush, and the ever-present smell of cod. But no Queens. The royal ladies were running farther south. Convoys grew larger and larger as a huge stockpile of war supplies was built up in Britain. U-boat attacks were now a rarity as our defences grew.

What Canadians call spring eventually made its chilly appearance and the fogs returned.

I recently read a book by a Canadian naval officer who had spent a considerable amount of time at sea. He had an intense dislike of shore establishments, doubtless with some reason, but I wish to take issue with him on one point. He implied that shore officers were happy in the security of their positions and tried to hold on to them. In all my time in both Ottawa and Halifax I can truthfully say that every officer I knew was doing his best to obtain a sea appointment or to be sent overseas.

I spent the winter trying to discover how a met. officer could go to sea. I found that the only possible way was to become a forecaster in a British aircraft carrier, but for that one had to pass the Royal Navy's course in meteorology.

I spent much time agitating to be sent to England for this purpose and was furious when, early in 1944, I discovered that a Canadian Wren officer had been sent ahead of me. I was all for the equality of women, but not *that* much equality. However, my constant bellyaching eventually paid off, and I was told that I could take the course at the Royal Naval College in Greenwich starting 22 May. I was to sail in a convoy leaving at the end of April from Halifax.

Janie returned to her home in Cleveland. I saw her off with mixed emotions.

As related earlier, I sailed as the only passenger in an interesting new type of vessel, a "MAC" ship, or merchant aircraft carrier. All their names began with "Mac." Ours was *Empire Macoll*. They were regular grain ships, but with a flight deck instead of the usual superstructure. Manned by British merchant-navy crews, each carried four small biplanes and a detachment of the Royal Navy's Fleet Air Arm – pilots and mechanics. The aircraft were called "Swordfish" officially, "Stringbags" unofficially. They looked like something that hadn't been quite good enough for World War I, but were beloved by all who knew them. In encounters with the

Germans they had often outwitted their opponents, who flew straight past them, never imagining that anything airborne could be that slow. They could also turn on a dime. Another endearing feature was their ability to survive a rough landing on a heaving deck. However, they *were* funny looking, appearing to consist largely of string. Several U-boat commanders had received nasty surprises on being attacked in mid-ocean by what seemed to be an apparition from the 1916 Western Front.

Our convoy HX289M consisted of 133 ships, the largest ever to sail up to that time.* From our position on the starboard quarter we could not see the entire convoy due to the immense area it covered. We went through the usual fog bank for two or three days, following a Norwegian tanker that trailed a fog buoy. This was a small homemade float carrying a flag of sorts. It was our captain's duty to keep his nose right up close to this contraption, even though often it was impossible to see the ship ahead.

My assigned bed was a bench in the wardroom, the social centre of the ship. Every night the young naval pilots mixed with the older sea dogs of the merchant navy. The flyers' conversation consisted exclusively of descriptions of adventures, aeronautical or biological, principally the latter. Their vocabulary, acquired in various less-than-elegant schools and enriched by their air training in Australia, was vivid, if monotonous.

When the fog cleared we turned around into the wind and flew off our aircraft. In due course they landed back on. This was a nerve-racking operation, as the flight deck was much smaller than those of the big fleet carriers and was rarely horizontal. It took cool judgment to avoid running over the edge.

We never saw any action because by now the Germans were aware that life for a U-boat in mid-Atlantic was becoming very dangerous. We had a blow or two from the north-west and rolled a good deal. Our most dangerous experiences were in deck-hockey games. This was played with a small square wooden puck and field-hockey sticks. No one wore life jackets, unlike in HMS *Leamington* where they had been compulsory. The games were wild and furious. Whenever the puck slid towards the edge of the

---

* I received this information in a letter from D. Ashby, Naval Historical Branch, 25 May 1989.

flight deck, which had no railing, these mad young Englishmen would race after it with wild abandon, regardless of the angle of the deck. They would put their brakes on just as they reached the tippy edge. I suppose they found this quite safe after landing aircraft on a rolling deck that looked like a postage stamp from the air.

One night the sky was clear and there was a full moon. It was an extraordinary sight. The ships, of course, showed no lights. Those to the north of us were a greyish, ghostly shade as they rolled along through the waves. But to the south the ocean glowed like a huge silver tray illuminated from below, one vast brilliant sheet under a midnight blue sky. The escort vessels were sharply silhouetted, jet black against the burnished sea.

When I wasn't watching the landings and takeoffs with moist palms or trying to keep from falling overboard in a deck-hockey game, I read a high-school history of England I had found in the wardroom. This greatly enhanced the sightseeing I did in the months ahead. I also helped the naval surgeon lieutenant who was assigned to decode the weather signals and draw up the maps. Once in mid-ocean we had a following wind of just the speed of the convoy, which was nine knots. There had been a strong gale between Greenland and Iceland some days previously, and huge swells from the storm rolled in incessantly from the north-west. The sky was dark grey, the sea almost black, but there were no whitecaps and no wind on our faces. The swells were the size of the ship. We rolled and rolled and rolled. Nothing stayed put, everything banged. However, it wasn't like a storm with spray and screaming wind. It was quite strange. I wasn't seasick then, or at any other time during the crossing, but I was glad when the swell subsided and we could walk from one point to another without hanging on for dear life.

When we were about two-thirds of the way across, the second mate, a gentle Scot, said to me quietly, "This is about where I was torpedoed for the third time." The only details I could get out of him were that he had been some hours in an open boat before being rescued.

After fourteen grey days the weather cleared. What seemed to be a low cloud came into sight on the horizon. This was Ireland. We steamed on, and the details of the Emerald Isle became clearer. It truly was green. The sky was broken, and shafts of sunlight shone down on the rugged cliffs and lush fields of the northwest coast. We were safe from U-boat attack now, and there was a mood of serenity as we all glided towards our destinations. As Ireland passed to the south on our starboard side, the convoy split into two halves. Ours turned south-east towards England and the others curved to the north-west, heading for Scotland. It took another day's sailing to reach the Mersey River, anchor off the great seaport of Liverpool, and embark on the story already told.

CHAPTER TWENTY

Gasometers?

As I lay in my hospital bed with these reminiscences of war running through my bandaged head, the real war was progressing without me. We followed the broadcasts daily, and in due course my eyesight improved and I could read again. Paris was liberated by the Americans and the French. The British were fighting through those terrible Flanders fields, and the Canadians were cleaning up the coast. Many V-1 launching sites had been captured but there were still enough left to fire a reasonably steady stream at England.

I had missed the last two weeks of the meteorology course and was slated to join the current session for its conclusion. However, I needed to regain my strength and had earned some leave. Orders came that I was to go back to the Canadian base at Greenock, Scotland, in order to head south again. At the time this seemed to be the navy moving in its mysterious way, but in due course I found that they were trying to evacuate all unessential personnel from the London area before the anticipated V-2 onslaught.

Sunday, 27 August, was my last full day in the hospital. VAD nurse Jean Crosthwaite and her radiographer friend Sheila organized a double date on their day off for me and a burned RAF pilot. We bicycled to a glen in the woods. The weather was perfect. Jean was a few years older than I and very pleasant. This was my first meeting with Sheila. She was my age, stunning, and very married. In talking to her I said something about the impossibility of resuming a career in music after such a long absence. She gave me hell and told me firmly not to let a little thing like a war interfere with my plans.

Both the advice and the advisor stuck in my mind.

On Monday, 28 August, I was deemed well enough to travel and went into London on my way to Scotland. There was very little air activity over the city that day, and we later found that out of ninety-four V-1s launched from the Continent, only four reached London. Two were sent spinning by the balloon cables, fighters shot down twenty-three, and sixty-five were destroyed in the air by the guns along the coast.

Our air forces were bombing the launching sites as soon as they were located, but they were hard to find. Over two thousand of our airmen lost their lives in these attacks. After the war it was learned that as many German launching-crew members were killed by premature explosions as by our raids. On that pleasant Monday, of course, we knew nothing of this, only that things were a bit quieter. The southern Englanders would have enjoyed a grim chuckle had they known how many doodlebugs blew up at the launching sites.

We also found out that our balloons, although stopping many V-1s, were having a bad time of it themselves. The Germans had installed wire cutters on the flying bombs and before the battle was over had cut loose or destroyed 630 balloons. However, by the end of August only one bomb in seven was reaching the London area – a great tribute to the defenders.

The navy thoughtfully provided me with a full-fledged first-class bedroom for the trip to Glasgow, a great improvement on my last mobile sleeping quarters, described earlier.

The next day I started my convalescent leave, a sight-seeing trip in a counterclockwise direction around England. I was already madly in love with Edinburgh and London and now added to my polytropolis amours such exquisite places as Worcester, Tewkesbury, Exeter (badly damaged), Salisbury, Winchester, Cambridge, York, and Durham. More towns were added to this list in the months to come.

On 6 September, Herbert Morrison, Minister of Home Security, had announced, "The Battle of London is won." This was a remarkable example of nonclairvoyance. The war may have been won, but it was most certainly not over, as the poor Londoners were to realize over the next seven months.

I had reached Cambridge on my grand tour about 8 September. It was completely undamaged, and I spent an afternoon walking around the peaceful town with its magnificent buildings and lawns. I hadn't heard sirens or seen any recent damage for ten days, and the war now seemed rather remote. But it returned in its noisy fashion during the night as the sirens screamed their hoarse warning. This alert was apparently caused by one of the piggy-back doodlebugs launched from over the North Sea. I felt that the Germans had tracked me down.

About the time that I was in Cambridge, unknown to most of the world, two strange events took place in London almost simultaneously. On 8 September, at 6:43 P.M. in Chiswick, northwest London, there was a sudden great explosion, followed by a roaring reverberation that gradually died away. Sixteen seconds later across the city in Epping, a similar inexplicable happening shook the district. Several people were killed and houses demolished. There had been no flying bombs for a few days, and there was no alert. Various guesses were made as to the reason for the explosions, the most common being gas leaks. In the next week eight more such mysterious explosions blasted various parts of the capital.

It was time for me to go back from Scotland to Greenwich to finish the course. I shared a day-type compartment on the overnight train with another naval officer who had recently come from London and was returning. He said that the V-2s had started. No announcement had been made. Rumours had spread that these ominous booms with their long reverbations were gasometers* blowing up, possibly due to sabotage. The officials did nothing to discourage such guesses as the enemy had no idea whether or not they were hitting the city. He said that the Londoners were now suspicious. There were too many gasometers going up in a short period. The sound was unlike anything heard before. A tremendous double explosion, followed by a gradual diminuendo dying away in a faint whisper. What kind of a bomb was this?

When I reached Greenwich I moved immediately into a dormitory in the naval college. I visited my friends from 42 Ashburnham Place, now living in a tiny basement apartment in a

---

* A gasometer is a large storage tank for natural gas, resembling a drum on its side.

house belonging to strangers who were not overly pleased at having been forced by the authorities to take in these poor homeless folk. Miss Hards, Billie, and George II were in fine form and welcomed me warmly. We chatted for some time. Everyone was pleased that the V-1s had stopped (or so we thought). Nobody knew much about the V-2s. The Anglican church that they attended had been destroyed, and the clergyman had gone to the country for a week or so of well-deserved rest, probably his first since 1939. Miss Hards, however, disapproved.

"'E couldn't tyke it. 'E ran aw'y," she said with scorn.

I also paid my respects to our old home. It was still standing, but it was dark inside and dangerous to enter. Out in the garden lying on its belly, neglected and forlorn, was our very own doodlebug. I was surprised at how small it was compared with even a fighter plane. Its nose had vanished, but the wings and fusilage were clearly recognizable. I broke off a piece of the light metal as a souvenir.

Beside the bomb was all that was left of the stout garden wall that had probably saved our lives. It consisted of a little ridge of broken bricks about an inch high. No other trace remained, except a few fragments in my part of the sitting room, as Miss Hards had told me. The Anderson shelter in which my ambulance-mates had been sleeping was dark and dank. The church and the houses beside it were ruined, and several other houses on the three streets showed considerable damage.

The instructors in the course also greeted me warmly, and I met a new set of classmates. We were now studying some purely nautical aspects of forecasting, such as wave action on a beach, and so on. However, I found the time rather unpleasant. I didn't seem to have quite thrown off an intestinal bug I'd acquired some time earlier, and every afternoon I felt a little seedy, a mood not helped by the new German habit of launching V-1s from aircraft soon after dark. These now approached from the north-east, which was at least a change. Also, from time to time, a distant "gasometer" could be heard exploding, although there had still been no announcement about them either by the British or the Germans.

Unlike the summer, the fall evenings became very dark soon after dinner. I now could see the full effect of the blackout. There was little traffic after sunset. Vehicles were required to mask their lights except for narrow slits. Traffic signals were also covered except for a small cross that could be seen only a few feet away. At some important intersections there were dim overhead lights carefully shielded from direct viewing by aircraft. There was a national shortage of light bulbs, and the British, normally a most honest race, became habitual criminals whenever they saw a public light bulb. Thus there were no bulbs left even in places where there were blackout curtains or painted windows, such as trains. However, once we had adapted to the dark we could see fairly well, but this took time, and we groped around blindly whenever we emerged from a bright room.

Life at the college itself was reasonably pleasant. Bath time was unexpectedly amusing. The tubs were long and deep, but early in the war in an effort to conserve fuel, orders had gone out that baths were not repeat not to be filled above a certain depth, six inches, I think. The Navy had ordered a line to be marked around the inside of all of His Majesty's bathtubs at the correct depth. Naturally, this was called the Plimsoll Line, after the mark around the hull of a ship indicating where she should ride in the water when fully loaded. However, despite the great elegance of our surroundings, there was not a bathtub plug in the entire building. Nor, in fact, in the entire kingdom. They were not being made and had all followed the light bulb into oblivion. But the row of bathtubs was just beside the billiard room. It was standard practice to help yourself to a ball and use it as a plug in the bathtub. They worked perfectly and had the advantage of rolling back into place if accidentally kicked. I sometimes wondered whether Henry VIII had had this problem in this his old palace. (Note to North American readers: such newfangled gadgets as showers were not yet in use.)

In mid-October I completed the course and was sent for practical experience to a naval air station at Henstridge in Somerset, on the Dorset border. I felt badly at leaving my friends still under attack. However, I was able to explore the lovely rural countryside in my free time. I had additional romances with Bath, Wells, Glastonbury, and Polperro. I was invited to spend a day or two at a most pleasant farm in Devon. While appreciating the hospitality, I felt I was in a foreign land. My hosts hardly seemed aware of what much of their country had suffered, nor did they have any particular knowledge of the V-2 attacks.

Regulations demanded that they sell all the cream they produced to the government for the armed forces, but they held back
a considerable amount for themselves. At tea they served the
famous Devonshire clotted cream, but I found it difficult to show
the enthusiasm they expected.

CHAPTER TWENTY-ONE

Supersonic Surprises

It was not until well into November that either the British or German governments made any mention of the V-2 attacks, which by then had been taking place for two months. This, it will be remembered, was in sharp contrast to the early stages of the V-1 onslaught. There had been a good deal of unofficial chatter about the new menace, and the weapon was described in articles allegedly from Swedish newspapers, but probably planted by the British. This was a technique sometimes used to keep the public subtly informed without official confirmation. When the details were announced, it was obvious that a new age had dawned.

Years earlier, one of the women who spent the war studying aerial photographs became suspicious of a certain long thin shadow on the ground at Peenemünde in the Baltic. This was an experimental station that had long been the subject of our surveillance and was the birthplace of the flying bomb. Our bombers had damaged it extensively on various occasions, though not to the point where it stopped work on this weapon, which was of course a rocket. The whole story of how the British obtained the specifications of the V-2 has been told in several places and is most intriguing. Suffice it to say that they knew a great deal more about the V-2s in advance than they had about the V-1s, which were a later development, although the first to be used against England.

The government had been extremely worried about how to defend against rockets, with good reason. These awesome weapons measured five feet in diameter (eleven across the fins), they were as tall as a four story building, and weighed 13.6 tons, which included a one-ton explosive warhead. Launched in Holland, they soared straight up in the air and gradually curved westwards, crossing the North Sea at a height of fifty to sixty *miles*! (Modern jet liners fly at a height of about five to seven miles.) The rockets travelled at the unbelievable speed of four thousand miles per hour. The trip from Holland took about three to four minutes. As their speed was faster than that of sound, the sound sequence was like running a tape recording backwards. The first thing the Londoners heard was a thunderlike crack as the sound barrier was broken, followed within a second by the explosion of the warhead. Next was heard the loud roar of the approach as it neared the ground, then the quieter sound from a little farther away, till finally the last thing heard was the very faint whisper made when the rocket first became audible. In short, the rocket beat its own sound. It also came too fast to be seen. The entire elaborate organization of air defence was useless. No alerts could be given as the V-2s could not be detected in advance. Radar operators, fighter pilots, the Royal Observer Corps, and gunners could only stand and wait, like the balloons. It is astonishing that the Nazis, never given to hiding their light under a bushel, delayed so long in announcing this technical miracle to their own people.

Hurtling down from a great height at that incredible speed, the V-2 had remarkable penetrating power, as can well be imagined. However, it did not cause anything like the lateral blast of either a doodlebug or the parachute mines of the Blitz.

Unlike the V-1s, the V-2s were not given a nickname by the British public. "Rocket" seemed good enough, especially when preceded by an adjective or two of one's own personal preference. A later age would call these deadly space monsters Intercontinental Ballistic Missiles, or ICBMS.

I saw where one had landed at Chelsea Hospital, a home for aged soldiers. The rocket made a deep crater several feet across and damaged buildings around it, but not to the extent of a V-1. Another V-2 landed next to the Westminster Bank in the high street of Greenwich, where I had an account. It knocked out a building and damaged the bank, but the walls remained standing. Still another destroyed a business a little farther along the street. Poor Greenwich certainly had a variety of unpleasant objects fall on it in five years of war!

On Saturday, 25 November, a tragedy occurred in the neighbouring borough of Deptford. The government had just relaxed its severe restrictions on the production of ice cream, and word spread through the area that a supply was being sold at Woolworth's. The store was filled with children, many of whom were tasting ice cream for the first time. A rocket struck the building, causing terrible carnage. One hundred and sixty people were killed instantly and 135 seriously wounded.

The Londoners debated among themselves which were worse, the V-1s or the V-2s. As during the Blitz, many believed that "If it's got your number on it, chum, you've 'ad it. It'll be all over before you know it." These people preferred the V-2s. The other school of thought, to which I belonged, pointed out that for every person killed there were several injured by flying debris. With a few seconds warning you could usually duck under something and at least get away from glass. We felt that the V-1s gave you a sporting chance.

After the advance notice we usually received when a doodlebug was approaching, it was an awesome thought that at any instant you might be blown to smithereens by a V-2 without even one second's warning. Often as I walked near windows I felt the urge to scurry past in case a rocket suddenly blew up in my vicinity.

Shortly before Christmas I was transferred to another naval air station in Somerset, this one at Yeovilton. There was still repair work to be done on my face, and during the winter and spring I made two or three further trips to East Grinstead. These were followed by short convalescent leaves, and then back to duty. I often visited London, attending concerts and plays, with an occasional trip to Greenwich.

On my short visits to the city I usually stayed at a most gracious centre of hospitality, the Liberal Club, which opened its doors to officers of the armed forces. Just off Whitehall near the Thames, its heyday had been in the latter part of the nineteenth century. Gladstone must have been a frequent visitor. Bombed and patched up, it was drab and dingy. The elderly staff were unfailingly courteous. They must have known every prominent liberal politician since the 1880s.

The building was tall by London standards and I was customarily assigned a room on the top floor, just under the roof. The plumbing in the bedrooms was almost as elaborate as at 42 Ashburnham Place, but possibly more user-friendly. Usually before I went to sleep I could hear the explosion of a V-2 or sometimes a

V-1 reverberating across the great darkened city. I would have preferred a little more between me and the sky. Nevertheless, I greatly appreciated the nostalgia trip back to the Victorian era, despite the occasional reminder of twentieth-century aerial technology.

I had heard a good deal about Picadilly. On this street the windows of many of London's classiest shops displayed by day what little elegance was left to buy in austere war-torn England. In the blackout, commercial enterprise of a different type took place.

When walking along the sidewalk, every few feet one could sense a girl lurking invisibly in the dark. They advertised their charms by scuffing their feet on the pavement, sometimes murmuring a silky "Good evening." Mindful of the desirability of keeping myself well informed as to current economic conditions, I occasionally asked the price. The answer was invariably "Two pounds." Supply and demand were obviously in balance.

Perhaps the most beautiful evening of my life took place in London in mid-winter. There had been a severe cold spell and several inches of snow had fallen. The temperature was well below freezing, the snow was light and dry, not slushy, and none of it had been swept or plowed. The numerous ledges and carvings of all the great buildings were etched in white, bringing out their contours in a way that would have delighted the architects. The moon was full. The traffic at night, as always, had died down to practically nothing and there were no other pedestrians. No lights were to be seen. My eyes became completely dark-adapted, so every detail was clearly visible in the crisp cold air. The snow under the midnight sky was a shimmering blue. Big Ben struck midnight, his tones strangely muffled. The Houses of Parliament and Westminster Abbey were wedding cakes frosted with layers of icing.

But once or twice while walking in fairyland I was brought back to reality when a distant grim boom sounded the death knell of more Londoners.

CHAPTER TWENTY-TWO

An Uproarious Met. Office

My new station, Yeovilton, was a permanent Fleet Air Arm base, unlike Henstridge, which was "for the duration only." We were training pilots on Seafires, Spitfires adapted for deck landings. The pilots were destined for carrier work in the Pacific. I was now considered a qualified watchkeeping met. officer by the Royal Navy.

The senior met. officer was Lieutenant Commander Peter Bevan, RNVR, a most fascinating man. A Cambridge graduate in science, he also held a law degree and specialized in patent law. A brilliant thinker and an inexhaustible source of information on many subjects, he was the wittiest person I have ever known. Another of the met. officers was Paulina Brandt, who had been in my class at Greenwich. Like Peter Bevan, she had a lively mind and was an excellent conversationalist. In the evenings the off-duty met. officers often returned to chat with the person on duty and generally a few drop-inners. I considered the met. office to be the social, intellectual, and cultural centre of Somerset.

Part of our popularity was because our rooms were warm, unlike practically everywhere else in the British Isles. We shared a squat cement building with the control tower, heated by the rarest artifact, a furnace. Several of us were on duty over Christmas. I had recently received from home a package of marshmallows, unknown in wartime Britain. To cheer ourselves up, I introduced the North American art of roasting marshmallows to my fellow duty types, using our large furnace in place of a campfire. We tied a fork to a mop handle and managed to produce some tasty burnt globules without melting the fork or setting fire to the mop. As usual after a marshmallow roast, we ended up sticky and feeling we'd overdone it.

One night when Peter Bevan was on leave we had a cold spell of considerable magnitude. The ground was snow covered, the sky was clear, and every twig sparkled with hoar frost. The temperature fell lower and lower until four different thermometers indicated 0° Farenheit (−18° Celcius). We were the coldest reporting station in the British Isles that night and were very proud of ourselves. One of our Wrens, coming to work late in the evening, actually froze an ear solid white. However, the district met. officer wouldn't believe our reports, and he phoned me for verification as I was acting commanding officer of the met. station. He put up quite an argument, and I nearly had to mail him her ear.

When Peter Bevan returned he was thrilled to hear of our notoriety. He kept the thermograph chart neatly folded in his wallet for years and proudly displayed it whenever he encountered another met. man.

One frosty morning I saw a strange sight when bicycling to work just before sunrise. The sky was cloudless, the visibility exceptionally good. There, suspended in the eastern sky, was a thin vertical cloud extending from the horizon to a great height, glowing a brilliant orange from the still hidden sun. It was like a straight neon tube. The duty met. and control-tower staffs had never seen anything like this phenomenon. As we studied it through binoculars it slowly became bent and eventually dissolved. We decided it was probably condensation from a V-2 as it hurtled down into London, over a hundred miles away. We even considered the possibility that the trail was made over Holland when the rocket was ascending. We never saw another, nor have we found any reference to vapour trails from v-2s.

Forecasting the weather in that area was a treacherous business. Whereas in Halifax, Nova Scotia, we had been plagued with the vagaries of the Fog God, in the southwest of England we had two principal enemies besides the Germans. One was instability. Gorgeous sunny mornings would soon deteriorate to glowering skies and, by noon, heavy showers interspersed with what we cheerily called "bright intervals." Commander Pack had taught us to remember on such mornings the old English maxim rendered in his accent as "It's too fine to lahst."

Our other problem was fast-moving low-pressure areas from the south-west. We received no surface observations from the sea, although one dogged aircraft made a hazardous flight most days to send back upper-air reports. Our most scientific technique was to keep a constant eye on the barometric pressure from the Scilly Isles, just west of Cornwall. When it started to fall we would shout a warning to the control tower so they could call back the student pilots, using the aeronautical equivalent of our old friend the Boats' Recall.

Senior officers often made short flights to nearby air stations, usually within fifty miles or so. Time and again the weather closed in before they could return and they were stuck there for two or three days. When the brass hats bothered to ask Peter Bevan for a forecast for such flights, which they did only rarely, he would quote what he called Bevan's Fourth Law: "It's quicker to bicycle." This proved to be true in an astonishing number of cases. (The first three Bevan Laws were never enunciated. When pressed, Peter would give no details beyond stating that numbers one and two related to molecular motion in gases and the third was obscene.) We strongly suspected that none of these actually existed.

Another reason for enjoying Yeovilton was an elderly gentleman named Collins. A civilian map plotter attached to our office, he had been in the Royal Navy most of his life and had climbed from the rank of Boy up three whole rates to Leading Seaman with a specialty in gunnery. He served in the navy through World War I, fighting at Jutland, and had spent some time at sea at the beginning of the current conflict until somebody saw his age and forced him to retire, much to his annoyance. He then joined the civil service and was a most useful member of our staff, partly because of his diligent work, and partly because of his skill as an entertainer. White of hair, red of face, humble of manner, he was respectful to all. He bicycled several miles to and from work each day, often arriving drenched to the skin, but never uttered a complaint. His language was as pure as the driven snow in normal conversation. But when he told stories, which was whenever we could get his mind off his work, all direct quotations were verbatim and unexpurgated. Our wall charts were scorched from authentic lower deckese.

One of our favourite stories featured a certain small and ancient vessel that had been commandeered to serve her country early in this war. All seaports in wartime have a ship stationed outside the harbour that challenges any vessel wishing to enter. If necessary, she sends boarding parties and examines the stranger's papers and cargo. At the start of the war, Collins, although trained for big-ship gunnery, was drafted to serve in this dubious craft, whose only weapon was a very old gun mounted on the fo'c'sle. It dated probably from the Boer War. Collins was in charge. They only had a few shells and were not permitted to waste any of them in test firings, so nobody knew whether or not the thing would work. Nevertheless, Collins trained his crew of ex-civilians with the diligence he would have shown if aboard HMS *Nelson*.

After months of routine patrol, a dirty old tramp steamer came chugging along, belching clouds of black smoke and heading for the harbour. The usual signals were made, but the captain was either asleep or drunk, or else didn't take the navy seriously. At any rate, he paid no attention and kept right on coming. Collins's captain, a very junior sub-lieutenant RNVR, ordered "Action Stations!" and the gun crew closed up smartly to their positions.

"Prepare to fire the main armament," shouted the captain.

"The myne armament!" Collins would interject editorially. "It was the only gun we 'ad!" At this point in the story, which was told countless times, Collins would be laughing so hard that tears ran down his glowing cheeks.

"Collins, aim to fire across her bows," bellowed the captain.

"Aye, aye, Sir," our hero replied.

The tramp continued on her perilous course.

"Fire!" screamed the captain.

There was a mighty bang and a cloud of smoke, but the shell only went about twenty-five yards before plopping harmlessly into the waves. The gun, having torn itself loose from its moorings, rolled over the side and into the sea with a mighty splash, never to be seen again. The sound woke up the errant captain, and he hove to. Thus a major naval battle was averted.

If beseeched long enough, Collins could occasionally be persuaded to sing us a song. He knew a great many, but only one can be printed. The peacetime navy, and a good many wartime sailors as well, had a low opinion of civilian dockyard workers, called mateys. They considered them overpaid and underworked. Collins told us how at the naval dockyard at Portsmouth, universally called Pompey by the navy, there was a bell at the entrance.

It was rung at 8:00 A.M. If a workman entered the gate after the bell had sounded, his pay was docked. The navy commemorated the scene with the following version of the hymn "Take My Life and Let It Be."

CAN A DOCKYARD MATEY RUN?
As rendered by Collins

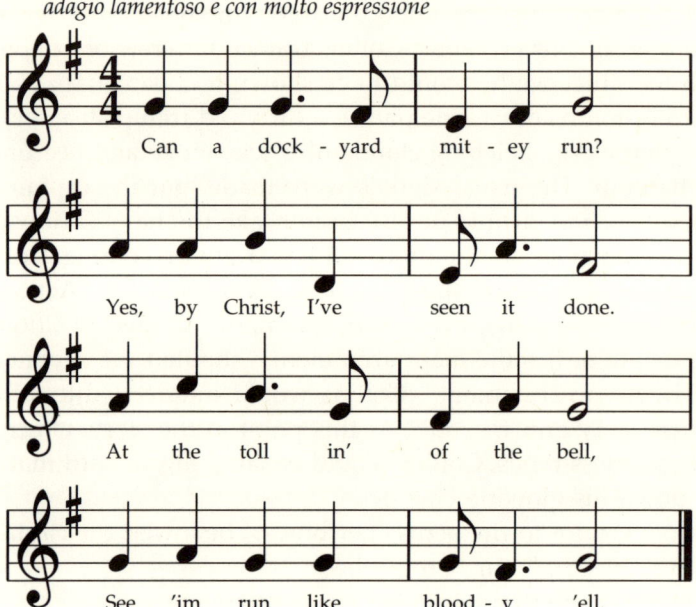

adagio lamentoso e con molto espressione

Can a dock - yard mit - ey run?

Yes, by Christ, I've seen it done.

At the toll in' of the bell,

See 'im run like blood - y 'ell.

I thoroughly enjoyed my time at Yeovilton. In addition to my splendid colleagues, the accommodations were good and warm, and the food was quite acceptable as wartime food went. Moreover, I had an ancient bicycle bearing the proud name "Royal Navy" on which I spent a good deal of my off-duty time exploring the lovely little hamlets in the neighbourhood. In many cases, their names ... Chiltern Cantello, Kington Magna – were longer than the main streets.

The control tower had a cement-floored hallway that always contained about a dozen parked bicycles, sheltering them from the frequent showers. Apparently this preyed on the mind of the chief flying instructor, a commander, RN. Eventually Daily Orders promulgated an edict to the effect that no repeat no bicycles were to be stowed in the control tower. The next day none were to be seen. But Peter Bevan used this as a text for a sermon on jurisprudence delivered in his customary humorously pontifical manner.

"It won't last. Just watch. It's a well-known principle of law that you can't legislate against the will of the majority."

After forty-eight hours it rained, and two or three bikes reappeared indoors. By the end of the week the hall had its full quota back. No mention was ever made again about the order.

In March I went back to East Grinstead to have my nose rebored.

McIndoe had finally been forced to take a long rest. Wing Commander Tilley was in the bizarre position of being an RCAF officer in charge of an English civilian hospital as well as both an RAF and an RCAF wing, and now he was the only plastic surgeon. He carried this enormous load with his customary unflappable charm.

The operation took place in the evening, the nose surgeon being a most eminent man who came to East Grinstead when required. I woke up from the anaesthetic in the middle of the night to find Wing Commander Tilley and the Canadian Matron at the foot of my bed. I had apparently been haemorrhaging and had swallowed a great deal of blood. It was characteristic of both of them that after a heavy day's work they still felt they could not leave me in that state and remained until the bleeding stopped.

Or at least temporarily. Thereafter I had small nosebleeds or blood clotting from a septum that had had too rough a life, what with one thing and another.

East Grinstead had received an influx of RCAF nurses who were efficient and pleasant, but I regret to say never seemed to develop the spirit of the English and Irish nurses or the Canadians who had been there during the Blitz. There was no sense of dedication and no apparent awareness that they were in a hospital where most of the patients had been disfigured in fiery crashes while flying in their defence.

I looked for Sheila, but she was home visiting her sick mother.

Back at Yeovilton, I continued my studies of the language, that is, British wartime slang. Little of this had crossed the ocean, and much of it is still unknown west of the Gulf Stream.

Of course, certain air talk was designed to confuse the Germans in dogfights and may even have done so for a day or two. Gibberish such as "bandits angels 15, tallyho!" was supposed to make the bewildered Messerschmidt pilots vainly fumble through their dictionaries as they hurtled through the blue. (Incidentally, the Fleet Air Arm claimed to have invented all this jargon and complained that the RAF had stolen it.)

The most significant contribution of WWII to North American speech was the new meaning of the verb "to have." "I've had it" meant "I'm fed up" in general (it could also be used specifically, as "I've *had* this pen," or "I've *had* George," but this usage never crossed the Atlantic).

Another useful survivor is "flap," a situation combining excitement with a touch of panic. The picture, of course, is that everyone is running around in circles cackling and flapping wings, as when a U-boat surfaces in the middle of a convoy or an admiral suddenly arrives for an unannounced inspection. A useful derivative is "unflappable."

"Type" never made it west, alas. Some scholars claimed it was derived from French. It simply meant a member of the human race, as in "I met this type in a bar." It didn't imply anything typical; "I met an army type" just meant a soldier.

The great disappointment is the demise of "prang," a brilliant invention. It showed remarkable versatility. As a noun it meant a crash or motor accident – "he was in a prang." It could be used as an intransitive verb – "he pranged," meaning "he crashed," – or transitively, meaning "to damage," as in "he pranged a wingtip," or even a fingernail. In special idiomatic usage it was a polite way of saying "he got her pregnant." What a pity "prang" has become as extinct as "flapper"! On the other hand, we can all rejoice that "whizzo" and "piss-poor" are buried in the rubble.

The wartime meals in Britain were meagre in quantity and generally poor in quality. Restaurant meals were on the whole adequate, but the rations for civilians were severely limited, especially meat. A family of four had about enough meat for one small helping once a week.

At Greenwich we ate comparatively well. The captain of the college firmly believed that the British naval officer was the highest form of life on this planet and should be fed accordingly. The meals were said to be the best in the navy. Today they would rate about the same as in a third-class American college cafeteria. At Henstridge and Yeovilton the meals were well below those at the college, but were still much better than the sparse and monotonous fare endured by civilians. Fresh eggs did not exist, but were replaced by a powder known as cardboard or rubber eggs. Sausages consisted of cellophanelike skins filled with breadcrumbs, pepper, and a dash of grease. Tea was plentiful but coffee was pale grey and tasted of formaldehyde. Vegetables were usually something called greens, a soggy mass of indeterminate origin. Desserts were often nice puddings or stewed fruit. Fresh fruit was rare. Mulligatawny soup appeared with relentless frequency. It was reputed to be made from leftover World War I mustard gas, but this could not be confirmed.

Afternoon tea was the best meal of the day, with cakes, bread, often jam, and lots of tea. We always left a meal feeling slightly hungry, and teatime was a most welcome break. Parcels of chocolate bars from home were also much appreciated.

On watch in the met. office, as elsewhere in the navy, we often had cocoa. His Majesty's official issue was suspiciously bitter. There was a widespread belief that it was heavily laced with saltpetre to reduce sexual activity. No such result was ever apparent.

A lovely English spring gradually developed. Two interesting aircraft made brief visits to our station. One was a Gloster Meteor, the first British jet we had ever seen. Although hard to believe, it could outrun a Spitfire by a large margin. The next unusual aircraft to pay us a visit was the navy's first helicopter. Several of us scrounged rides, one passenger at a time. It was a strange sensation to fly low over our hangars, sometimes stopping in mid-air, sometimes moving sideways, with the ground visible between my knees through the low plastic window. When we returned, the pilot hovered about four feet above the ground and suggested that I jump out. I did so, with no greater consequence than getting my hair blown awry by the downdraft of the rotor blades. This impressed the senior officers who were watching, as nobody had ever seen anyone jump out of an aircraft in flight without a parachute. It made an enormous impression on all of us, and I predicted that after the war private helicopters would be as common as cars. Fortunately, or unfortunately, at the time of writing this prophecy has yet to come true.

Incidentally, Yeovilton later became the main training school for naval helicopter pilots and included among its distinguished pupils the royal princes. Graduates from this station made a major contribution to victory in the Falkland Islands.

During the winter and spring of 1944–45, southwestern England and Wales were free from aerial attacks. The southeast was, as usual, hit intermittently by V-1s and V-2s. Occasionally the Germans directed a few V-1s against northern England. On Christmas Eve fifty piggy-backed flying bombs were aimed at Manchester, one of which reached the city while seventeen fell outside. Thirty-seven people were killed and sixty-seven injured.

The Belgian port of Antwerp was devastated by flying bombs and rockets after it had been captured by our armies. The Canadians moved up the coast of Holland and finally seized what proved to be the last of the launching sites. However, the tension continued through April because no one knew for sure it was all over.

Germany was invaded, the Rhine was crossed in many places, and the world was appalled at the discovery of the concentration camps. The Soviets were moving towards Berlin, and the RAF bombed Potsdam at the end of April in their support. Some have criticized these late raids, including the bombing of Dresden. But at the time the shockwaves generated by the pictures of the extermination ovens, the concentration camps, and the emaciated creatures found inside their gates created an intense desire to crush once and for all the machine that produced these atrocities.

During my fourteen months in the UK we heard very little about the Battle of the Atlantic. For the previous two years I had seen the convoy plot every working day and knew each triumph and tragedy. Now, for all I knew the ocean might have dried up. After the war we found that towards the end things had become very unpleasant again. The Germans had invented the schnorkel, which enabled the U-boats to use their big diesel engines while submerged and to recharge their batteries while presenting only a tiny target for our radar. A few new submarines were being developed that, when submerged, could outrun our standard escorts. They had also formed the unpleasant habit of releasing fake conning towers and our aircraft wasted a good many depth charges on these dummies. Much of the new action took place near the British Isles. Thus the submarine war ended in much the same location in which it had started in WWI.

We followed the great battles in the Pacific with increasing interest. The student pilots we were training were being groomed for carrier duty after the European war ended, and many of us expected to go with them. One sunny spring day was spoiled by a film that the Americans had made featuring kamikaze attacks on carriers. It spared no details. War films on the whole were not always viewed with favour by the men in uniform, but this was received with local critical acclaim. The accolade bestowed on it by most of the young pilots was "no bullshit." They tried to cheer themselves up with a grim chuckle by saying, "They sweep up the fingernails after each showing."

The atomic bomb saved us from seeing the real thing.

CHAPTER TWENTY-THREE

Hot Wars End, Cold War Begins

In May, I returned to East Grinstead for another operation.

Sheila was on temporary duty at another hospital.

The London area had been quiet since the end of March. On 7 September 1944, Duncan Sandys, who held the august title of chairman, War Cabinet Committee for Defence against Flying Bombs and Rockets, had issued another of those famous statements that come back to haunt the stater. He had said that the Battle of London was over "except for a last few shots." The V-2s started the next day, and in the months ahead hundreds of flying bombs and rockets spread destruction among the inhabitants of England.

The V-1s launched from aircraft had made their unwelcome debut on 9 July, but the main onslaught was in the fall months, when around six hundred were sighted. They stopped after 14 January, and the defenders had a respite until 5 March, when a new type started arriving. These were longer range, and launched from hurriedly built sites in Holland. One hundred and four came over in March but the defenders shot down eighty-one. The last V-1 exploded near Sittingbourne, Kent, on 29 March 1945.

The V-2s continued all winter. In February, 116 struck London, followed by 115 in March. On 27 March the last V-2 landed in Kynasten Road, Orpington, which, incidentally, on 11 September had been the first place in Kent to be hit by a rocket.

The Canadian army's advances in Holland captured launching sites and cut supply lines, thus ending the pilotless attacks. To the public, these historic last explosions were just more bangs. No one could be sure that all the launching sites had been accounted for, nor that there was no new secret weapon. The V-3s had been in the backs of our minds for a year, and in view of past false forecasts we were highly sceptical of assurances that there would be no more attacks. (The V-3s were supposed to be missiles launched from large ramps, firing multiple weapons, but fortunately these were never developed.)

The Allied armies were near triumph, although we were afraid that even if the German government surrendered, diehard members of the armed forces would continue to fight, especially in Norway or the mountainous regions of Germany. We particularly worried about pilots staging a last-ditch air raid.

I was to have an eyebrow graft on Wednesday, 9 May. On Monday 7 May, negotiations for the German surrender were in full swing and we hoped for an announcement at any minute. Time went on while we held our breaths. At 3:00 P.M. on Tuesday, 8 May 1945, Churchill announced that Germany had surrendered unconditionally. Our relief knew no bounds, although we kept our ears open in case of an air attack.

There is still ambiguity as to the date of the long-hoped-for Victory in Europe, or VE, Day. Some people still say that it was Tuesday, but Churchill announced that the next day, Wednesday, 9 May, would be the official celebration of peace and a general holiday for all. There would have been little point in declaring Tuesday a holiday at 3:00 P.M. when most people were at work.

All surgical operations at East Grinstead for Wednesday, 9 May, were cancelled, so VAD Jean Crosthwaite and I took a train to London. We had tea with friends of hers and then I went on my way.

All during my growing-up years I had been determined to be in London for the next Armistice Day, having missed the one in 1918. I don't know why it fascinated me so. I had heard many descriptions of the unbounded joy that exploded when that terrible conflict came to an end. So I was thrilled that circumstances permitted me to fulfill this great ambition.

It was a perfect spring day with warm sunshine and a pale blue sky. The trees were covered with tiny green leaves. I wandered through the crowds in a happy state and eventually ended up in the well-known area in front of Buckingham Palace. Tens of thousands of people were there, hoping that the royal family would come out on the balcony and wave. Everyone was smiling, chatting amiably with friends or strangers, and casting an occasional glance at the empty balcony. There was none of the drunken revelry that I had heard about at the time of the first armistice, although that may have occurred later in the evening. I kept an ear out in case some fanatical Nazi had one last V-1 left over and wanted to spoil the fun, but none came. In fact, the surrender proved total and complete, accomplished in an orderly manner. They had had enough. We stood and stood, and every so often we'd chant at the top of our lungs "We want the King!" I kept looking at my watch because I knew I had to catch the last train back to East Grinstead. Throughout the war, commuter trains stopped running early in the evening, and there was obviously nowhere to stay in crowded London that night. Time went on and still nobody appeared on the famous balcony. I calculated how long it would take me to reach Victoria Station and waited as long as I dared. Finally I left, running most of the way, without having seen the king or Churchill. I had to be content with their pictures in the paper the next morning.

The train to East Grinstead passed through miles of those humble homes that back onto the railroad tracks. Bomb damage was everywhere. Groups of happy people were bringing armfuls of kindling from wrecked houses, and already bonfires were being lit in the narrow streets. This is a centuries-old British Isles custom for special celebrations. One of the most famous in historic times was when Charles II returned to the throne in 1660. The tradition still survived.

Night had fallen by the time I reached the hospital. I found a charming lady – I don't remember whether she was a nurse, a Wren, or a WAAF. We climbed a hill from which we could look across the fields to the north. There in the black night was a golden glow filling the whole northern sky. These were the bonfires of London. Such a sight had been seen all too often during the Blitz, but tonight it was caused by a populace released from the fear of death striking from the skies. It was a double relief for those whose menfolk no longer faced the risk of death in Europe.

My new friend and I strolled back to the hospital together in a thoughtful but happy silence. As was so often the case in wartime, we never saw each other again.

My operation took place on Friday. Hair from my temple was sewn on where my eyebrow once had been, the one that was tossed into the garbage can at the Greenwich Hospital. On the same day Reg Hyde, one of the worst-burned pilots, was to be given a new nose. He had been at East Grinstead since the Battle of Britain in 1940. They gave him his anaesthetic, but when they began to operate, his chest had raised a rash from the disinfectant applied just before the operation and they couldn't make him a nose at that time. He and I were the only anaesthetized cases that day, and it interested me that we convalesced at the same speed, even though I had had surgery and he had not.

On Sunday evening, 13 May, a kindly old gentleman took us both for a drive in the placid countryside. This had been his practice for some years, one of many kindnesses showered on us by the residents of East Grinstead. He was allowed a small amount of petrol for this purpose. We entered a pub, and, as usual, no one seemed to notice our grotesque appearance. The others had a drink and I had my customary ginger beer.

Churchill came on the radio to make a speech, thanking the nation for its sacrifices. However, he sent a chill through us all with a reminder that we could not consider freedom safe as long as there were police states in the world, and that we could not rest until Japan had been defeated. This was really the first of his "Iron Curtain warnings," long before Fulton, Missouri. We returned to the hospital with our euphoric mood badly damaged.

How right Churchill was!

Britain had been under intermittent air attack for ten of the last thirty years. The final phase, by unmanned weapons, had lasted for over nine months. It had caused a great deal of tension, sorrow, dislocation, and damage, but did not significantly help the German cause. On the contrary, it may have increased the Allied desire and determination to destroy German cities.

The official postwar statistics indicate that 8,564 flying bombs were launched against England from the early morning of Tuesday, 13 June 1944, to Thursday, 29 March 1945. Of these, 3,957 were destroyed as follows: 1,847 (forty-six percent) by aircraft; 1,866 (forty-seven percent) by guns; 232 (five percent) by balloons; and 12 by the navy. The problems of inventing new weapons can be seen from the fact that 1,693 were defective and caused no harm to our side, often killing the launchers.

The casualty figures for V-1s were placed at 6,139 killed and 17,239 seriously injured.

In the period from 8 September 1944, to 27 March 1945, 1,115 V-2 rockets reached England, 518 of which hit London. In all, V-2s killed 2,855 people and badly injured 6,268.

Thus the combined V-1 and V-2 attack killed 8,994, mostly civilians, and seriously injured 23,507.* Ninety-two percent of fatal casualties occurred in London. When added to the 51,601 civilians and service people who died in conventional raids, the sombre death toll from air attacks in Great Britain was 60,595. These figures of course are not comparable to those for the atomic bombs, or even for the great fire storms on the German cities, but they do represent a tremendous strain spread over a long period.

The Nazi minister of Production, Albert Speer, complained that the rockets were uneconomical. One V-2 took as long to make as five fighter planes and cost as much as twenty V-1s. On the other hand, they did impress on all of us the enormous technical skill of the German war machine. And they gave the world a foretaste of what World War III could be like.

A vast number of houses and businesses were destroyed or rendered unusable. The southern and eastern suburbs of London suffered the most. Croydon was one of the worst; 142 flying bombs hit the borough, the most for a single day being eight. Over a thousand houses were completely destroyed and 57,000 badly damaged. These two figures add up to more than the total number of houses in the borough, the explanation being that many houses were hit, repaired, and hit again.

Penge, a southeastern suburb, holds the dubious distinction of having received the highest bomb count per acre. Every house in the borough was destroyed or damaged, two-thirds having been hit more than once. Greenwich was among the leaders, but while it was in second place for a while during the summer, others nosed it out later, notably Wandsworth, Lewisham, Camberwell, and Woolwich. There is a certain ambiguity in the figures as some

---

* These figures are from H. Saunders *Royal Air Force, 1939–1945*, vol. 3, 175. P.G. Cooksley, in *The Flying Bomb*, cites 8,958 and 24,504 respectively.

include V-2s, while others are based on bombs per acre or casualties.

By the end of September 1944, over a million houses had been damaged by the new weapons together with 149 schools, 111 churches, and 98 hospitals. By January 1945, 130,000 men were at work in London repairing buildings. Forty-five thousand were brought in from outside, most of whom had to sleep in temporary accommodations, further adding to the housing problem.

The British cabinet had early resolved not to allow the V-1s to detract from the main military thrust through France. When this decision was reversed a vast amount of manpower and material was brought to bear on the new menace. Two thousand Allied airmen lost their lives attacking launching sites and supply depots, and a huge number of men and women were manning the guns, balloons, and rescue operations. As an illustration of the tremendous burden placed on the population, it can be stated that nearly one million homeless people and Civil Defence workers were fed after raids by mobile canteens of the Women's Voluntary Services. This great organization also distributed clothing to the destitute and conducted house-to-house sweeps after incidents, recording casualties and providing comfort. Many of these women were killed while on duty.

One shudders to contemplate what the results might have been if the double attack had started a year earlier and if the Germans had been able to produce their weapons in greater numbers. We were not in a position at that time to overrun the launching sites and our air forces were not as large as later.

In the 1930s much was said and written about the ability of air power alone to win wars. Since then, conventional wisdom downgrades air attacks on civilians, claiming that in WWII it was indecisive and merely served to stiffen civilian morale. I doubt whether these views are held by people who spent any time in London. If the attack had been much heavier, it could well have changed the outcome. It is also my opinion that the reason why German militarism appears to be dead is that in WWII, unlike in the earlier conflict, the Germans saw the devastating effects of war on their own cities.

In view of the success of the early pilotless weapons, I continue to be amazed fifty years later at the continuing emphasis on piloted bombers, which are to me as obsolete as cavalry.

My new eyebrow took, and I returned to Yeovilton for a few weeks of lovely spring weather and an atmosphere of temporary peace. During the winter an order had come from Ottawa for me to proceed to Newfoundland to become staff officer (Met.) at St John's. I was furious. It meant no sea duty and also would have put me in a position inferior to the one I had held for a year in Halifax, despite the fact that I was now much better qualified. I had sent out a firm signal indicating that it was essential for me to remain in the UK until my plastic surgery was completed. This quieted them down temporarily while I continued to plot how to get to sea.

With the European war over now, the weak-kneed government under Mackenzie King took an extraordinary position: they gave all of us in the temporary armed forces the opportunity either to resign or continue the war against Japan. We had signed on "for the duration" and to many thousands of us it was unthinkable to quit while we were still at war. It also seemed cowardly to let the United States and Great Britain carry on alone against the Japanese. I volunteered to serve in the Pacific and was delighted when in due course I was appointed met. officer in a new Canadian light fleet carrier, HMCS *Magnificent*. She was building in Belfast, but would not be ready for some time. All of us Pacific types were given immediate priority to go home on leave before taking up our new duties. I made one last trip to East Grinstead, collected my medical records, and said goodbye to Joan Ricketts, "Sis" Mealey, Jean Crosthwaite, and other nurses and patients I knew.

Sheila was in the darkroom and couldn't be reached.

On passing through London I said goodbye to Miss Hards and my former housemates, and travelled north to Greenock. While waiting for our sailing date, I went into Glasgow to a concert given by the great violonist Yehudi Menuhin. Towards the end, Menuhin, who was then in his mid-twenties, invited an elderly Scot up on the stage from the audience. It was the beloved comedian-singer Sir Harry Lauder, long retired. He and the violinist chatted amiably while we all eavesdropped and then Menuhin persuaded Sir Harry to sing one of his favourite songs. The old pro's ability to spellbind an audience was undiminished. As he and Menuhin hugged each other after the song, we in the audience didn't know whether to laugh or cry, so we just cheered our heads off.

In early July hundreds of us sailed in a Canadian escort carrier, every nook and cranny of which was filled with double-decker bunks. A large contingent of Canadian Wrens was sailing with us. This was the first time any of us had seen women aboard a warship. During the seven days of placid sailing they went around in convoys, presumably to protect their virtue. We couldn't even get within whistling distance.

On reaching Halifax harbour, we tied up at the same pier that had always been used by the Queens. There was a crowd on the dock, composed mostly of Wrens from the local offices who had been ordered to take the afternoon off to welcome us. As I came down the gangplank, a girl whom I had seen many times at the dockyard office rushed up and embraced me across the railing. Hers was the first "old" face I had seen for many months.

In the harbour were two German U-boats. They had been operating off Nova Scotia when the war ended and had come to Halifax to surrender. I was shown over one. It was astonishing how small they looked from the outside and how cramped inside. Even boarding them on the surface was a trifle claustrophobic. It brought home the courage, skill, and determination of the men who had spent the war in such uncomfortable and dangerous vessels. It was hard to believe that eighty or so of these tiny craft nearly swung the balance in Hitler's favour.

In due course I arrived one evening at the Toronto railway station. A large crowd, held back by brass rails, was waiting to greet returning service people from our train. As I hove in sight a vision in white slipped between the rails and hurled herself on me. It was the greatest moment of my life.

In a few seconds I was also hugging my mother and shaking hands with Uncle Reg, resplendent in the uniform of the first commodore in the Royal Canadian Naval Volunteer Reserve.

Several days later, Janie and I started on a trip we had dreamed about ever since we had fallen in love eight years earlier at the University of Toronto. I wanted to show her the beauty of western Canada. We took a passenger ship through the Great Lakes, Georgian Bay being considerably rougher than anything I had experienced on my recent Atlantic crossing.

Mother joined us on the West Coast, and the three of us spent a few days in Victoria at a small hotel. Most of the clientele were ladies of elegant upbringing but reduced income. At breakfast on

the morning of Monday, 6 August, a permanent resident at a
nearby table was reading a newspaper propped up against a vase
of flowers. The silence of the dining-room was briefly interrupted
as she read out to no one in particular, "The Americans seem to
have dropped some new type of large bomb on Japan." It had
landed on a city I had never heard of, but it sounded like Hero-
something.

I was not unduly impressed, as American journalists were not
noted for their understatement. I assumed they had built a bomb
comparable to the British "Ten Ton Tess," also called the "Tall
Boy," which our Lancaster bombers had been using against concrete U-boat pens and also the battleship *Tirpitz*.

There was talk of an ultimatum having been issued to the
Japanese, but this was nothing new. Everyone anticipated two or
three more years of war, fighting island by island, and then invading Japan itself.

We had made reservations to take the steamer up the coast to
Prince Rupert and were to sail on Saturday night. That morning
in Vancouver I decided that I had better attend to a matter that
had been on my mind for some time – tropical uniforms. We went
to the local branch of a national tailoring company that had made
my regular uniforms in the past. The salesman was a young man
my own age. I told him I wanted to order five white tropical
uniforms.

"We can't get the cloth."

"Well, where *can* I get uniforms?"

"Search me."

"Don't you know there's still a war on?" I asked, getting hot
under the collar.

"That's your problem," he said, and walked off.

By now I was blistering. His attitude reflected far too much of
what I had seen since returning from devastated England to plush
Canada, particularly on the West Coast. I decided I would have
to suspend my search until we returned from Prince Rupert. I
didn't much like the idea of fighting the Japanese in my
underwear.

We sailed that night, Saturday, 11 August. On Tuesday, 14 August, we spent the day at Ocean Falls while our ship was loading
newsprint. The falls were not vertical, but a series of streams
flowing down a rocky slope with smooth bare slabs between the

little pools and cascades. It was a lovely afternoon. Janie and I climbed up the rocks, jumping over the little rivulets. There were two or three other small groups of people in the distance, but otherwise we were alone. We sat down and enjoyed the view. Even though I was in uniform we took off our shoes and dabbled our feet in the clear sparkling water of a nearby streamlet.

Suddenly the whistle on the paper mill let out a series of shrieks. It sounded like a fire signal. We looked but could see no flames or smoke. Just then our ship's whistle joined in the uproar. A couple on the rocks some distance away started to wave wildly. It was rather alarming.

At last the great truth dawned – peace! Japan must have surrendered! We shouted and waved and embraced. It was a glorious few minutes, with the two whistles screaming and the distant shouts and waves now from a number of people and the knowledge that, much to our surprise, the war must be over.

Then the whistles stopped. Emotions became strangely mixed. For the last twelve years we had been either anticipating or fighting a war. Now, for the first time in nearly six years, we were at peace.

There I was, twenty-seven years of age, no war, no carrier, no plans, no job, no civilian qualifications. And I didn't even have my shoes on.

But I had a wonderful wife. And I didn't have to buy those white uniforms!

CHAPTER TWENTY-FOUR

Have We Forgotten?

1949

The old troopship *Aquitania* swam slowly into Southampton harbour. She was back in passenger service on what was called an austerity basis and had only a few more crossings left in her before heading to the scrap heap. Her wartime grey was painted over with her peacetime colors, black hull and white superstructure, funnels brilliant red with black tops, the hallmark of all Cunarders.

I was thrilled to be bringing Janie back to see the island and the people that I had come to love so dearly. After the war, it had been impossible for casual tourists to obtain transatlantic passage in either ships or aircraft. However, by July 1949, we were able to book a berth in this gallant old vessel sailing from Halifax. As we made our way through the maze of cranes and ships to tie up astern of the luxury liner *Caronia*, I once more saw what had so moved me five years earlier on approaching Liverpool – masts protruding from the sea. Several pairs of these nautical tombstones marked the graves of drowned ships and drowned men.

The rest of the port was in good shape, having received a high priority for repairs. We took the fast train through the English countryside, and soon I was showing Janie around Central London. The covering over the statue of Eros in Piccadilly Circus had been removed, showing him in all his naked splendour. Otherwise, there was little visible change. The bombsites were even grimier than before and the sooty buildings even sootier. The government was placing all its emphasis on the reconstruction of essential buildings and houses rather than beauty treatments. The huge devastated area north-east of St Paul's was untouched, and the Thames was lined with the blackened shells of burnt-out warehouses.

We stayed a few days with Peter Bevan and his wife Phyllis and gave them one of two hams that we had brought over. Meat rationing was still in effect, and no Britisher had seen a complete ham for ten years. It was interesting to note that despite Peter's high-level connections, he had no more to eat or to wear than Miss Hards and her friends in the East End.

There were many similarities between the England of 1949 and the Soviet Union we saw during a visit in 1980. The governments of both countries had a desperate need for hard currency, which meant US dollars, and took every step to encourage tourists and to give them special benefits. There were still shortages of most consumer goods, and queues formed at shops whenever goodies came in for sale.

We had tea with the Coggans at their suburban home and were delighted to renew our acquaintance with this wonderful family. The future archbishop of Canterbury expressed his gratification at seeing that the knot he had tied was still secure.

We had a most pleasant stay with ex-VAD nurse Jean Crosthwaite and her father on their lovely estate in the country west of London. They took us to lunch at the Royal Henley Yacht Club. The dining-room was filled with the gentry of England. Part way through the meal I embarrassed myself by being stung in the leg by some devastating creature. Within a few seconds I was tingling all over. I felt quite lightheaded. I went to the men's room and had a thorough search but couldn't find the culprit. There was a tiny spot on the side of my calf. My host and hostess were of course mortified that there was an insect loose at the Royal Henley Yacht Club. They also considered it a bit much that in addition to my having been bombed in their country, I was now also stung. I began to wonder whether it had all been my imagination, but for several weeks I carried as proof a large red welt that hurt whenever I went near it. Jean gave me news of several of our East Grinstead friends. She herself was soon to be married.

Sheila was living happily ever after.

Janie and I made a memorable trip to Greenwich. She was appalled at the amount of damage she could see from the railway

as we journeyed out to the eastern suburb. There were rows and
rows of roofless shells, once homes of the poor. Interspersed were
burnt-out factories, warehouses, churches. All looked just as they
had in 1945.

The high street of Greenwich also was unchanged. Janie remem-
bers the blackened interiors, the blasted walls, the stained
wallpaper and plaster, and the gaps in the façades where buildings no longer existed.

We visited Miss Hards and her two housemates, Billie and
George II. They were still in the basement apartment where they
had been placed "temporarily" after we were bombed. Miss
Hards had suffered a great heartbreak. The British government
had organized all the rebuilding of houses. Nobody could outbid
his neighbour for the repair crews that were hard at work all over
the country. You took your turn. The plan was to repair first those
houses that had suffered little damage and could be made into
useful habitations with little work – broken windows repaired, a
hole in the roof fixed, etc. Then came those buildings that required
more extensive repairs, and, last of all, houses to be rebuilt completely from the ground up. This way a much larger number of
families could be rehabilitated sooner. One day Miss Hards had
been returning from shopping and passed the end of Ashburnham Place on her way to her present home. She had looked along
that street many a time and had seen the house's solid outline
standing securely in place, though dark inside. It was one of the
middle category and she had been expecting the workers to show
up soon.

As she walked along on this occasion she saw vehicles and
workmen at her house and, after a glance, was horrified to see
that it was half demolished. They were knocking down the walls
as she hurried up to them and said, "What's all this? Why are you
tearing my house down?"

"Them's our orders, madam," replied the foreman, showing her
an official-looking document. She took one glance at it and said,
"But this is for 42 Ashburnham *Grove*. This is 42 Ashburnham
Place."

When the workmen took a look at the street sign, they apologized and stopped work, leaving the house in a half-demolished
condition. Years later, we found that she had never been able to
move back into her own beloved house. It was too low on the

priority list to be completely rebuilt, and she passed away before it was done.

Our three friends were as pleasant as ever, though naturally a little older. But it was sad to find them in a tiny living space with its underlying feeling of defeat and hopelessness.

We gave them the second ham, for which they were touchingly grateful. It was extraordinary how something that was so common and readily available to us could mean so much to these long-suffering friends of ours. Miss Hards served tea, handing me a cup for the last time. The china was of cheap quality, her mother's best set having been destroyed during the war.

After tea, I showed Janie a number of the famous and less-famous places in Greenwich, from the Royal Observatory and the naval college to the curbstone outside the men's room where I had flattened myself that unforgettable Saturday night. We also examined what was left of 42 Ashburnham Place, so unnecessarily destroyed. We then spent four weeks driving around Britain, using special tourist petrol coupons. My favourite towns were as enchanting as ever. But it struck us that, apart from a few plaques in churches dedicated to individuals, we saw no memorials to civilians.

1954, 1958, 1960, 1964, 1968, 1969

On repeated visits to Britain we could see gradual recovery from the air raids. The area north-east of St Paul's sprouted isolated skyscrapers of great height and unrelieved monotony. There appeared to be no attempt at planning – huge office buildings took shape beside dusty lots, empty basements, and ruined churches. Sooty façades were sandblasted and took on a lightness that had not been seen for two hundred years. The authorities had erected a standard prefabricated temporary house on the lot at Ashburnham Place. It was one of a vast number built to solve the housing problem. They were plain but comfortable, with modern conveniences.

Our old friends had passed away.

An imposing monument on the Embankment honoured the dead of the Royal Air Force. A section of the Imperial War Museum gave a splendid depiction of the air raids. But around the city we still saw no memorials to civilians, nor were there any

sites being preserved for future generations. Pamphlets for tourists listed innumerable tours that could be taken, some on foot,
some by bus. These seemed to cater to every taste, from night-
club *habitués* and history lovers to fans of Dickens or Jack the
Ripper. But none concerned the Blitz nor the V-bombs. Visitors
were not being told of acts of heroism and sacrifice on *this* battlefield, where children and women ranked with men as soldiers of
freedom.

Guidebooks large and small continued to ignore the great
events that took place in the two latest wars on British soil.
Whether a thin leaflet about a parish church or a weighty tome
published by an internationally known firm, the authors seemed
more interested in some skirmish during the Wars of the Roses
than the great battles fought by civilians, many of whom were
still living. The name of the bishop who authorized alterations to
a west transept is preserved forever, but no reference is made to
the women and men who risked their lives night after night as
firewatchers on the steep roof of the cathedral. The grave of a
knight who fought in the Crusades is carefully tended, but what
about the rescue worker who was killed when a wall collapsed?

1972

We visited the battlefields of World War I. On Vimy Ridge we
were shown over the trenches preserved as part of the war memorial to the Canadian forces that captured those bloodstained
mudholes in 1917. The lawn surrounding the great memorial was
bumpy with shell craters. Signs in French and English cautioned
against walking on the lawns or in the woods because of unexploded shells. A day or two before we arrived, two French youths
had ignored these warnings and were killed when a shell blew
up after having been buried for some fifty-five years. The beautiful sorrowing stone face of the woman on the monument gazed
across the dreary plains to the east, now mourning for two more
young lives taken by that terrible war.

On Saturday, 3 June, we drove towards the Normandy beaches.
We first came to Pegasus Bridge, named after the emblem on the
shoulder patches of the British 6th Airborne Division. They had
captured the bridge after landing by gliders a few minutes past
midnight on 6 June 1944. This was a vital communication link if
the invading forces were to move eastward.

We had read detailed accounts of this gallant operation and were deeply moved to be at such a site. We drove across the bridge and stopped in front of a small café, the upper half of which was the home of the owners. A sign in French said it was the first house in Europe to be liberated by the Allied forces. It gave the time as a few minutes after 11:00 P.M. on 5 June; this was puzzling as our historians always refer to D-Day as 6 June. The explanation is that the army used British double summer time while the Continent was one hour behind.

We were alone in the parking lot in front of the café mulling this over when suddenly a car careened up, slammed on its brakes, skidded to a stop, and disgorged four burly men in British camouflage battledress. One seemed to be the leader. He walked around a few steps, apparently selecting a site for something. They were talking English with a Cockney accent.

"'Ow about right 'ere, mites," said the obvious leader.

He pointed to a place in the parking lot a few yards away. They discussed it, and the others seemed to agree. Finally, my curiosity could stand it no longer and I went up and said, "May I ask what you gentlemen are doing here?"

"We're settin' up a memorial 'ere," one of them said.

We chatted and they told us they were the advance party of the 6th Airborne Division, which was holding a reunion at nearby Caen commemorating the twenty-eight anniversary of their landing on D-Day.

"We were enormously impressed in '44 when we heard about you fellows on the radio," I said. "Thank you, gentlemen, for what you did for all of us."

"That's right nice of you, sir. Chaps, has anybody ever said that to us before?" the burliest asked. They shook their heads.

"Come on in and let us buy you a drink," one suggested.

The soldiers practically carried us inside and sat us down at a long scrubbed wooden table. They insisted on buying our soft drinks while they had something stronger. One paratrooper had already been celebrating to a considerable extent and took little further interest in the proceedings.

The leader told us that the middle-aged lady who was the bartender had grown up in the house and lived there during the entire four years of German occupation.

"Imagine! Think of the courage *that* took!" he said with awe in his voice.

She had made a collection of weapons after D-Day and in her backroom was a veritable arsenal.

"You nime it, she's got it," one soldier interjected admiringly.

In due course they had to return to Caen, and we parted after this heart-warming encounter.

The whole experience was dreamlike. It was as though we had been walking on the battlefield of Waterloo and a few of the Duke of Wellington's soldiers had dropped by for a pint.

The next day we toured the Normandy beaches. They were dotted with memorabilia – plaques describing various events, cemeteries, a memorial tank, German artillery encasements, and, in the village square of Le Hamel, a service attended by veterans of the Dorsetshire Regiment and local residents.

We saw no memorials to civilians.

1975

Malta had been a key base for the British during the war and had been savagely bombed several times a day for months on end. The king awarded the George Cross, the highest decoration for civilian gallantry, to the entire island in recognition of the fortitude of its inhabitants.

Our taxi driver took us through Valletta, the capital, which had been almost annihilated. The building stone taken from the local quarries is easily cut and soon weathers to a remarkable hardness when exposed to the air. This city had been so skilfully rebuilt that there was no trace of damage. Hence we were startled when we rounded a corner and came upon one devastated building. All that was left was a rubble-filled basement and a few columns. It contrasted starkly to the rest of the city.

"What was that building?" I asked the driver.

"The Opera House," was the reply.

"Why didn't they rebuild it?" I asked.

"Who needs opera?"

We saw no memorials to civilians.

1978-79

We explored the World War I battlefields again, this time over New Year's. First, the harbour of Zeebrugge, Belgium, where in

1918 the British navy had deliberately sunk a cement-filled block-ship to seal in several U-boats that were in the harbour.

We walked out on the mole, the long breakwater extending into the North Sea to protect the harbour. The wind was cold and raw, the sea grey and rough. A plaque marked where HMS *Vindictive* had tied up and landed her assault party. The atmosphere was thoroughly suited to the chilling story.

We drove west to Nieuport and turned south, following what had been for four gruesome years the old front line. Except for one ruined church tower there was virtually no indication of war in the green fields until we reached St Julien, just outside Ypres. This was where the Canadians withstood the first gas attack in 1915 and later was the scene of much carnage. A cement block-house beside the road added a grim touch to the otherwise peaceful rural scene.

We entered Ypres in the late afternoon. It was completely rebuilt and resembled a medieval town. The only evidence that it had been the centre of terrible fighting in World War I was provided by a few shell holes in the old ramparts and a broken wall near the new cathedral. We took a room in a simple little hotel.

That evening we attended the nightly ceremony at the massive Menin Gate memorial. This is a covered bridge across the old moat surrounding the town. It was really more like a tunnel some twenty-five or thirty yards long. All over the inside walls were carved the names of British servicemen who died fighting in that area and whose bodies had never been found. The ghastly total was about 55,000 "Missing in Action."

Shortly before the ceremony began the police stopped all traffic. A little group of us stood around shivering. Puddles were frozen, and the wind moaned through the tunnel. At the announced time, a man in a long civilian overcoat arrived, took out a trumpet, and played the British "Last Post." The mournful but noble melody reverberated through the sepulchral acoustics as we all stood rigidly at attention. It was perhaps the most moving musical performance I have experienced in a life devoted to musical per-formances. The player was from the local fire brigade. This ceremony has been held every night since the end of World War I.

The trumpeter left, the traffic resumed, and the frozen specta-tors began to chat quietly. A very short old man with a pronounced Scottish burr started up a conversation with a young lady near him. A German, she was deeply touched by the ceremony. They had a cordial and friendly talk. After she left I spoke to him. His brother had been killed there and he himself had done a good deal of fighting in the nearby trenches. He was back once more to visit the scene of his nightmares.

"The trouble is," he said, "we were fighting the wrong enemy. We should have been fighting the French. They're verry derrrty people. Why, sometimes their toilets are arranged so that women have to walk right by the urinals!"

When we woke up the next morning it was to a white world, just turning pink as the sun rose in a hazy blue sky. Several inches of dry snow had fallen in the night, and the gothic façades were incredibly beautiful. Every little ledge was outlined in white. We drove to Hill 62, where trenches were preserved. Nobody was around. As was usual in that sector, there was water at the bottom of the snow-covered trenches. They looked miserable. A blasted tree trunk, a survivor of Sanctuary Wood, brought to mind the terrible pictures of the battlefields. We drove on white roads to Passchendaele, which was totally rebuilt, and back to Ypres. After lunch we found our car wouldn't move because slush had accumulated around the wheels and then frozen solid. As it was New Year's Eve, no one was available to tow the car and thaw it out.

Next morning it was even colder. A kind Flemish gentleman helped us break the ice, literally, and we were on our way.

Outside the town we saw a number of blockhouses and other strong points deep in snow. On the infamous patch of ground known as Hill 60, perhaps the bloodiest acre in Europe, fresh tracks showed where a happy child had pulled his sled across the shell craters. The monument to the British dead of WWI had been vandalized by the Germans in 1940.

Every two or three miles there was a British cemetery, tenderly maintained by the Commonwealth War Graves' Commission. Headstones stood silently in sombre rows. A shelter contained a book giving the locations of the graves and describing the fighting that had taken place on that ground. There was also a visitors' book. It was touching to see how many people had travelled long distances to visit these hallowed places, some within the previous few days.

In certain cemeteries there were only a few dozen headstones to mark the graves of thousands of bodies. The unmarked graves were of soldiers who could not be identified. These were some of the "Missing in Action."

Outside St Eloi a cemetery contained the usual number of WWI graves, but in addition there were a few of the Britons killed in late May 1940, men who never reached Dunkirk. One was a twenty-four-year-old bandsman of the Royal Scots Fusiliers.

On the windswept Messines Ridge we found Lone Tree Crater, now a deserted frozen lake half the size of a football field, with steep sides. This was where the British, after months of silent mining, had blown a huge hole in the German front line. The explosion could be heard in England. The brotherly organization Toc H had purchased the land to preserve it as a memorial to the fallen.

There do not seem to be organizations buying real estate in London to preserve the memory of its defenders.*

We continued past "Plugstreet" to Armentières, France.

Next day it was snowing heavily on Vimy Ridge. Visibility was about a hundred metres. As we drove from the north we saw no car tracks nor any other traces of humanity until we reached the parking area. A lone small car sat there accumulating a white cover. Footsteps led to the trenches, where we met a young Australian couple shivering on their summer holidays. They were both experts on WWI, and showed us a fascinating book on the subject, *Before Endeavors Fade*, by Rose Coombs. It described virtually every surviving object of interest on the British front – monuments, dugouts, trenches, gun emplacements, etc., together with details of related actions.

---

* In Amsterdam in 1960 it was almost impossible to locate the house where Anne Frank and her family and friends had hidden. It was not on any lists of places to visit, and inquiries produced only shrugs. Finally a taxi driver took us there. The door was locked and the blinds drawn. On a faded piece of cardboard in the window was roughly lettered *Anne Frank Huis*. Sometime later the Anne Frank Foundation bought the property and opened the hideaway to the public. It has become almost a shrine for visitors from everywhere, with young women predominating. As at Dachau concentration camp we were greatly impressed by the number of Germans who seemed deeply moved by the experience.

We have not found a similar guidebook on the Battle of London. All traces of that great struggle are being obliterated by bulldozers and concrete.

The afternoon was clear and sunny, the countryside a dazzling white. Outside Albert a section of trenches has been preserved to honour the Newfoundland Regiment, which suffered grievously there in the Battle of the Somme, 1916. It is maintained by the Canadian government. We drove up a road in what was once No Man's Land. The snow had drifted and soon we were hopelessly stuck. However, two men and a boy were trudging up the hill and gave us a push. We talked French back and forth until finally I heard the man speak in plain old Canadian English to his son, who replied in same. We burst out laughing as we realized we were all Canadians. He was the permanent caretaker of the memorial. On reaching the top we risked going no further and drove back to Arras with the setting sun making the snow-covered fields glow orange and pink. Looking at this beautiful winter scene it was hard to imagine that here the British lost sixty thousand men in one day and fought for six months for a few miles of mud. This was the Somme.

In the museum at Arras there was a small but fascinating section devoted to the Resistance in wwii. On a wall in the former moat of the fort were a number of plaques dedicated to citizens executed there during the occupation.

At last there was some recognition of civilian gallantry and sacrifice.

We headed towards Brussels from the south. On the highway just north-east of Mons, Belgium, was a memorial marking where the advance cavalry of the British Expeditionary Force had fired the first shots of wwi at the Germans on 22 August 1914. By coincidence, it was also the exact spot reached by the Canadian army on the last day of the war, 11 November 1918.

A few miles farther, in Nivelles, a plaque on an innocent-looking house honoured the memory of civilians who had been tortured and murdered in that building when it was the local Gestapo headquarters.

In mid-afternoon we reached Waterloo. No effort had been spared to immortalize *that* battle. The much-fought-over farm of La Haye Sainte was a tangible reminder of the action.

Why is so little preserved in London to remind future generations that the great city, like Waterloo, was also the scene of heroic action and selfless devotion to duty?

1980

This was the year of the Great Nosebleed. To appreciate its historical significance and connection with WWII, a good deal of background briefing is necessary.

The college where I conducted for forty-one and a half years supplied all the formal music for the 1980 Winter Olympics, held ninety miles away at Lake Placid, New York. I trained our large chorus to perform at the opening ceremonies under our orchestral conductor, Richard Stephan, and conducted the chorus and symphony orchestra in the ecumenical service and also the closing ceremonies. My small choir also appeared, opening the meeting of the International Olympic Committee (IOC) before the games started. This was the occasion when Secretary of State Cyrus Vance startled the world by announcing that the US would withdraw from the Summer Olympics unless the Soviet army pulled out of Afghanistan.

An account I wrote at the time runs as follows:

The IOC meeting started when we sang the *Olympic Hymn* – no announcement, we just started.

It was an odd day. Police dogs sniffing every seat beforehand, security men thick, and the ground-floor window of the ladies' room wide open and unguarded! Nobody asked me for a pass once I put my tails on. I could have gone anywhere. We all walked across the frozen lake for dinner, a lovely ten minutes in the cold, clear air – twilight one way, dark with sparkling lights for the return.

The opening ceremonies and both rehearsals were perishing. The temperature was 8 degrees F (−13° C). I had to hold Dick Stephan's music down because of the wind and turn pages and literally brush off snowflakes. I also had to massage the frozen foot of the first cellist (as I always say, one percent of conducting is conducting). I had a walkie-talkie with a poor earphone held in my ear with a bandaid (referred to later as the Official Olympic Band-Aid). I had to pick out the cues for the orchestra from all the other fascinating chatter on the air – "Czechoslovakia's in the wrong place – tell them to move over … Tommie, I haven't picked up Governor Carey yet … Fire One…Fire the aerial bombs! … FIRE ONE! … someone go and tell them to fire … oh well, never mind, it's too late. Get Greece moving … GET GREECE MOVING … No, NOT everyone, just Greece, THEN Albania … Brock, tell Dick to end at the next cadence … Brock, BROCK, stop the orchestra." All of this almost inaudible due to Bob Washburn's great march. I was a wreck afterwards and frozen stiff.

The ecumenical service was held in the arena and featured the original Great Darkness, straight out of *Israel in Egypt*. Just as we started, someone turned off almost all the lights. We were bathed in what my clerical father used to refer to as Deep Religious Gloom. The congregation couldn't read the words they were supposed to sing. Nor could the performers see the music. Finally we got some lights turned on by refusing to continue.

My piece, *Let the Spirit Soar*, had been selected by the committee, but neither of the two electronic interludes came over the loudspeakers. Twice I waited for a century and then continued. A solo prayer in the middle never got prayed, I know not why. But a Russian lady told a band member a week later that this piece was the highlight of the service. I later asked the technician why he hadn't played the tape. He said that no one had told him when to do it. I said there were full instructions enclosed with the tape. He said he wasn't supposed to read instructions, only to follow cues.

But the 7,500 people had a great time and even clapped 2 Corinthians for the first time since it was written.

The closing ceremonies with their one and only rehearsal were like a surrealist movie. It would take months to write it all. But just a few highlights …

We arrived for our rehearsal Saturday and found that the IOC had forbidden us to put the orchestra platforms, chairs, or music stands on the ice due to the skating competition that evening. So we stacked up the 118-piece orchestra in clumps on the bleachers, with a large entrance in the middle and the chorus going up to heaven (half the 450 singers were in the second balcony). The soundmen failed to show, as usual, so we could hardly hear ourselves. I conducted standing on the ice near the blue line. Actually, it went very well.

There were several other groups taking part. None of us had been told by the famous Hollywood producer what the others were supposed to be doing. We rehearsed the entrance of the athletes, with placards for each country and flagbearers, but no athletes. I was conducting Bill

Maul's "Parade of the Nations" and was getting along nicely, conducting from memory and watching the performers, when I saw movement out of the corner of my eye. Here was Greece, coming between me and the performers! The other nations followed. The whole ruddy lot, complete with flags, passed between me and the orchestra. Janie finally convinced the producer that the parade was going through what would eventually be the orchestra, and he rerouted it. We had to do it all again.

The performance was even weirder. I still can't believe it all happened. Due to the two hockey games and the skating exhibition no one could get in to rehearse that day. At 7:30 I was given the latest "final" script, about twelve pages long, written that afternoon. THEN the changes started coming! There was hardly a page that didn't have a major revision scribbled on it after 7:30.

Chuck Mangione's band was testing the PA with ear-splitting shrieks, and our trusty bullhorn was all but ineffective as I tried to tell the orchestra about the new instructions, such as Don't do the repeats we worked out in the Maul march, percussion do eight bars of 4'4 cadence and then we repeat "El Capitan" in tempo as they'll still be marching, but watch for a sudden ending when they tell me to stop ... "Blue Danube" has to be during the intermission, but kill all repeats, but not *da capos*. Add a bass drumroll to the end of the Olympic Hymn if the flame hasn't been extinguished, but not otherwise ... add a snare roll between "The Star-Spangled Banner" and the Yugoslavian national anthem – etc., etc ... All these new orders were passed back from stand to stand in increasingly corrupt form.

Then we started.

The light man from the ecumenical service got to work again and down we went to almost total darkness. I conducted "The Star-Spangled Banner" by radar, and the Fort Ticonderoga Fife and Drum Corps marched around invisibly. Then suddenly the full hockey lights came on and stayed on all night. I found out later that there had been a great fight on the intercom, presumably as the TV people killed my old enemy, the light man.

Then the show went on, Greece didn't get lost until just before the end, nobody marched through the orchestra, Don Curry and Dorothy Hamill skated their bits, Chuck Mangione's pieces ended, much to my surprise. Lord Killanin made a nice speech, we whistled "Colonel Bogey," and did Elliot del Borgo's "When Dreams are Dreamed" cut from eleven minutes to three to allow for TV commercials. We sang the Greek and Yugoslavian national anthems and they got the right flags up, thereby delaying the start of WWIII a little longer.

The space under the stands had been converted into a "green room." A few easy chairs made the bare surroundings a trifle cozier. I had been on my feet for about three hours, so I sank into the soft upholstery with a great sense of relief. I was still in my tails, with white tie and hard-boiled white shirt.

Minor nosebleeds had been bothering me occasionally ever since my flying bomb. I had been afraid that one would start on the podium. I pictured myself turning around to bow, looking as though I had just been assassinated. Fortunately my nose, realizing the international significance of the occasion, comported itself with impeccable dignity.

But only while we were on world television. As soon as I settled in the chair my nose seized the opportunity to make a nuisance of itself and started to bleed.

One of the great characteristics of the Olympic Games is the free intermingling of different nationalities in various support positions. On this occasion the duty medical officer and nurse were both German. They spotted my trouble and immediately rallied round. The medical officer was about my age, handsome and courteous. The nurse was young and dazzling. Ice was brought and she held one hand while the doctor took my pulse on the other side. He said it seemed a trifle fast. I was surprised it wasn't one long roar.

"Do you often have nosebleeds?" he asked in slightly accented English. With some awkwardness I assured him not to worry, that it would soon stop.

"How long have you had these?" he inquired solicitously.

"Um, well, since one of your V-1s hit me in 1944," I stammered.

"Oh, I'm so sorry," he apologized with obvious concern.

We had a good laugh. The nosebleed soon stopped. I thanked them both and said, "Now we're even."

As far as I was concerned, World War II ended at that moment.

1981

In June we again visited the World War I battlefields, this time with much better driving conditions. We were given a private tour of the Newfoundland memorial trenches. Between Amiens and Arras little other evidence could be seen of the war. Dotted here and there beside the roads were monuments to various British military units that had served in that area, erected lovingly by the surviving soldiers in memory of their fallen comrades.

The cemeteries now displayed neat green lawns and flower-beds. There were no caretakers, and no evidence of vandalism. In 1989, I met a young Canadian whose mother was Dutch and had lived in Holland during World War II. He had recently bicycled through the battlefields and spent every night sleeping in one of these desolate outposts of empire, hoping to recapture the feeling of World War I.

This haunting land still exerts a hypnotic effect on many people. Rose Coombs, a WAAF plotter during the Battle of Britain, has visited here over a hundred times, often leading tours.

Apart from the cenotaphs, the cemeteries, and an occasional water-filled crater, there was little evidence that the bloodiest war in history had been fought in these same fields sixty-five or so years earlier. Crops were neatly tended, damage had been totally repaired, and many new buildings stood where nothing but terror had reigned for four horrible years.

I visited two of the three locations where my Uncle Arch had been wounded. One was at Wancourt, the other at St Eloi, where several huge craters, filled with water, gaped skywards. The Canadian Second Division, including his outfit, the 27th Battalion, City of Winnipeg, had fought here in appalling conditions. One crater had a small diving board extending over its muddy depths. I shuddered at the thought of what might be encountered on plunging into that murky hell-hole.

We visited the junkyard at Passchendaele. It consisted of mountains of rusting scrap metal such as old car radiators, but it also held remnants of World War I, still being dug up by the local farmers. A truck had just driven in. Included in its cargo was a dangerous collection of six or eight unexploded artillery shells, recently unearthed. They sat in a neat row, looking very lethal.

I bought a battered British helmet, destined to repose in state at home beside the German helmet we had found near Arras in 1926 when my father was visiting graves of his parishioners. This had been my boyhood pride and joy and my mother's *bête noire*. Muddy and rusty, a two-inch gash showed where some sharp object had been driven through it. We boys preferred to think it was a shell splinter that finished off the wearer, but some kill-joy adults pooh-poohed that idea, claiming it had more likely been made by a plow after the war.

Another pathetic sight in the junkyard was a trombone, green with corrosion and covered with caked mud, its slide bent almost at right angles. What had happened to its player? Was he lying under one of those headstones? Was his name carved inside the Menin Gate? Or was he enjoying his retirement in England or Germany in some home for octogenarian musicians?

Could he have been the young Scots bandsman who never reached Dunkirk?

1982–83

On New Year's Eve, 1982, we attended a performance in London at the National Opera. Afterwards we stood for a while on the steps of St Martin's in the Fields and watched the huge throng gathering in Trafalgar Square to greet the new year. No taxis were available so we walked for an hour back to our hotel, meeting hordes of young people heading to the square. Many were drinking as they passed.

During the night we were puzzled at the number of ambulances we heard dashing past. It was as though there had been an air raid. At breakfast next morning we read that a terrible tragedy had taken place in Trafalgar Square after we left. When the new year had been welcomed and the crowd dispersed, a number of people were found lying on the pavement. Two or three were dead and many others seriously injured. There was no actual violence – they had all simply been crushed by the huge crowd.

After breakfast we took a walk through the centre of the financial district, the City. It was deserted. The day was cold and grey, and we still felt shocked by the events of the night before. We cut through the area that had received the worst of the bombing and had once been the vast wilderness of basements described earlier. It was now dense with new buildings, including the magnificent Museum of the City of London, and other commercial developments. Everything was higgledy-piggledy, with no breathing space left.

Suddenly we came across an anachronism – a relic of the eighteenth century in the form of an old brewery. Its sign showed it to be Whitbread's. We stood outside looking at this old brick

building, the only pre-war structure for blocks around. On its wall
was a plaque that read as follows:

THEIR MAJESTIES
KING GEORGE III & QUEEN CHARLOTTE
WERE RECEIVED IN THIS BREWERY
BY
SAMUEL WHITBREAD
14TH MAY 1787

An entrance suitable for horse-drawn wagons led into the old
courtyard. As we curiously gazed inside, a well-dressed gentleman came out and spotted us.

"Happy New Year!" he said cordially in an impeccable accent.
"Would you like to come in and have a look around?"

We accepted his invitation and he showed us through the empty
brewery.

"How is it that this place survived the Blitz when everything
was destroyed?" I asked.

"Our employees had a great loyalty to this company," he
explained. "They set up a very efficient fire-watchers system and
took turns on the roof all through the Blitz, putting out any
incendiary bombs that came our way. You see, most of the damage
in this area was caused by fire. Our brave staff pulled us through
it largely unscathed."

He gave us an illustrated pamphlet relating the history of the
brewery. It had been owned and operated by the same family for
over two hundred years, as shown by the plaque.

On parting, I thanked him and asked for his name.

"Oh, I'm the current Whitbread," he replied casually.

We didn't have the heart to tell him that we were both lifelong
teetotallers!

1988

London was humming. It was enjoying a tremendous boom, with
little unemployment and a great deal of building activity.

The underground rooms from which Churchill and the other
leaders ran the war were at last open to the public. They were
fascinating. The only concession to human comfort was an

occasional electric heater, totally inadequate against a penetrating English winter. Nor did the building appear to be particularly safe. A direct hit by a large bomb in the entranceway might have changed the course of history.

Otherwise, London seemed to have forgotten the war. Although a few chunks of masonry were still missing from the façades of the famous buildings along Whitehall, there were virtually no plaques or other evidence of the air attacks of either WWI or WWII. Younger visitors to the capital would have no reminders of the anguish the great city and its great citizens had endured.

The war memorial at East Grinstead lists only forty-eight air-raid victims, although at least 108 and more likely 123 were killed in one raid alone, that of 9 July 1943.

We took the boat trip from the Houses of Parliament down-stream to Greenwich. Many new buildings lined the shores, erected on land formerly occupied by burned-out warehouses. Beside Tower Bridge, once a scene of devastation, an expensive hotel had been built. Some of the highest-priced condominiums on this planet face the historic waterway. Vast new factories and other commercial buildings were rising in the area that had suf-fered so severely in the raids of September 1940. Cranes by the dozen crowded the skyline once dotted with barrage balloons. Even the Isle of Dogs was coming to life again.

We landed in Greenwich. The naval college and observatory looked as magnificent as ever. A splendid exhibit was on display commemorating the four hundredth anniversary of the Spanish Armada. A fine new house stood on the lot at 42 Ashburnham Place. The spot where our V-1 had landed was now a peaceful courtyard. The Methodist church next door had been rebuilt, as had the house around the corner formerly occupied by my ambulance-mates. The courtyard where I had hurried to Green-wich's first flying-bomb incident was neatly rebuilt with a new wooden fence. Merryweather's Fire Engine Factory retained its old walls but was converted into offices. The North Pole pub was its usual busy self. The public toilet under the traffic island against which the longshoreman John and I had sheltered from the doo-dlebug had been replaced by a flowerbed, greatly improving the atmosphere.

There were other new buildings in the high street, although some gaps had not been filled, including the one beside my old bank. I asked the teller, "Do you know what caused that vacant lot next door?"

"No, sir, I don't."

"A V-2 rocket landed there. Your bank was closed for quite a while," I told him.

"Is that so, sir?" he said politely and went on counting his money.

Together with many other tourists from various parts of the world, we climbed aboard the boat and headed back to Westminster. The guide talked on and on with his historical facts and historical stories about everyone from Judge Jeffreys to Anne Boleyn. He made jokes about the prices of the apartments and the foreign ownership of the shopping plazas. But not a word to indicate that we were sailing through one of the great battlefields of history, the winning of which preserved civilization as we know it.

By the end of the trip I couldn't contain my exasperation any longer. I said to him, "Why don't you tell these people something about the Blitz or the V-bombs, how all that area until a short time ago was covered with gutted warehouses, ruined docks, and roofless houses?"

"Oh, that's old stuff. Who cares about it now?"

1989

East Germany was prosperous and tightly controlled. J.S. Bach's bones, or what are believed to be his, have stopped their wandering and rest securely beneath the chancel floor of his old church, St Thomas's, in Leipzig. An English choir and a German orchestra combined to perform his B Minor Mass, a most encouraging sign of mutual forgiveness.

In East Berlin, vast areas of bombed buildings had been replaced by plazas, lawns, flowerbeds, pools, and fountains. Ultramodern skyscrapers lined up symmetrically. There was none of the hodgepodge of Central London and there was plenty of breathing space.

Beside the Wall, near the Brandenburg Gate, a small entrance to a concrete bunker could be seen in a rubbly construction site. The black hole symbolized the evil it once contained. It was Hitler's deathplace.

The government-run tour took us to a cemetery dedicated to Soviet soldiers killed while capturing the city. We were not shown any graves of German soldiers. On many downtown buildings were signs of what London had been spared – scars from rifle bullets. These were the only memorials to the civilian population.

However, in Dresden, near palaces that were being skilfully restored, a ruined church was left in its damaged condition as a memorial to the vast number of civilians killed in the firestorm.

Back in London we visited Kew Gardens. The exceptional heat-wave that lasted through much of June had scorched the smooth lawns to straw. We crowded aboard a river boat for the return trip. About a hundred Londoners and tourists basked in the eighty-five-degree sunshine enjoying the tranquil cruise. The engine of this elderly vessel overheated and had to be given an occasional rest.

As we disembarked we noticed a small bronze oval-shaped plaque on a bulkhead. It said that this valiant ship had served at Dunkirk, forty-nine years earlier. No one else bothered to read it.

We spent a weekend at Southend-on-Sea. On Sunday afternoon the village and railway station were crowded with Londoners cooling off from the city's unusual heat. It must have resembled the scene on 13 August 1917, just before the Gothas struck. The local museum made no mention of that horrible disaster, surely the most significant event in this resort's history.

A monstrous cement block had been preserved on the espla-nade, an antitank device built in 1940 for the invasion that never came. But we saw no memorial to civilians.

On Tuesday morning, 13 June, we headed for Bethnal Green to find the site where London's first flying bomb had landed, forty-five years ago that morning. One newspaper briefly mentioned the anniversary, but none of the others we saw thought it worth a line.

We arrived at the Underground station, which had been the scene of a terrible tragedy during the Blitz. On asking for Grove Road we were told that it was one stop farther, in Mile End. We set out on foot in the heat. This district of London is just across the Thames from Greenwich and its neighbours, close to the Isle of Dogs. After a few inquiries, some of which sent us by mistake to Globe Road, we reached Grove Road. An elderly lady was waiting for the bus.

"Excuse me, but can you please tell me in which direction is the railway bridge where the first flying bomb landed?" I inquired.

"Right along there, luv, where you see the arch. You'll see a plaque that'll tell you about it." She pointed to a railway overpass a quarter of a mile away.

We walked eagerly south, and sure enough, on a grubby railroad bridge was a round disk on which was printed:

ENGLISH HERITAGE
THE FIRST FLYING BOMB ON LONDON FELL HERE
13 JUNE 1944

None of the passers-by gave the plaque a glance.

This was the LNER line that was reported blocked on that historic morning. No actual damage was visible, but the difference between Victorian and forty-five-year-old brickwork could be clearly seen. Beside the south-east corner of the bridge were some pleasant postwar houses, obviously replacing those where six people had been killed and forty-two injured that fateful dawn. This was the mysterious aircraft that had flown over the Royal Naval College and caused so much speculation.

Inquiries produced the information that the plaque had been erected through the persistence of a local resident, Joseph V. Waters.

At last we had found a memorial commemorating the air war!

1990

On leaving the site of the memorial plaque in 1989, we had popped into a small shop to ask the way to the nearest tube station. In chatting with the owner of the neon-sign business, we reminded him that it was forty-five years to the day that the first flying bomb had hit the bridge down the road.

"Yes, I suppose it is," he replied.

He then told us that the gentleman who organized the mounting of the plaque, Mr Joseph V. Waters, was a friend of his. We went on our way.

On reaching our hotel I realized that I had foolishly not learned more about Mr Waters. By looking in the phonebook I managed to find the address of the neon-sign store. Once back home, I wrote to the owner and asked him to pass on to Mr Waters a letter in which I requested more details about the plaque and congratulated him on his efforts. This produced a winter-long correspondence on the subject of air raids and Victorian East London, on which he was an authority. To our astonishment we found that his home was one of those destroyed by the first flying bomb. In June we visited Mr Waters and he took us to see the site. The house was rebuilt and occupied, one of several in a row. The house next door was still an empty shell, open to the sky, filled with rubble and weeds. Joseph had been away in the navy when the bomb fell. Family members had been injured, though none fatally. He took us on a fascinating walk around the area. We first visited the touching angel monument to the victims of the 1917 bomb that fell on the North-Street School, Poplar. We concluded with a tour of the Ragged School Museum, of which he is a founding member. Housed in a Victorian school building beside an ancient canal, the exhibit has a wonderful atmosphere.

Mr Waters is one person who respects the struggles and triumphs of the past and energetically works to preserve its memories.

1994

The fiftieth anniversary of D-Day was given lavish recognition. For weeks beforehand television was filled with excellent programs featuring that tremendous achievement. On 6 June itself, Normandy was jammed with celebrities and veterans. But in North America I never caught a single reference to the great battle that had started a week later, and to which this volume pays tribute.

2000 PLUS

Why *should* there be memorials for civilians? Why *should* warden's command posts, fire halls, first-aid stations, barrage-balloon moorings, anti-aircraft gun emplacements, and other reminders of the air raids be preserved? Why should the navy, army, and air force have commemorated their battles with monuments and other memorabilia of heroism and sacrifice? Why not forget it all?

Because the world forgot it all in the 1930s and allowed evil to rise again unchecked. As a result, fifty million people were killed, two-thirds of them civilians.

With a new century approaching, the younger generation must be reminded of the awesome dangers always present in this world. The causes of war must be studied. Constant vigilance must be maintained to detect the development of these causes. Our best brains must be devoted to safeguarding us from the resultant horrors. We all must strive our utmost to prevent wars large and small, whether by forming a true world government with an international police force, or by breaking up large nations into small independent states, or by arms limitations, disarmament, or developing powerful armed forces; by redrawing national borders, or fortifying them, or dissolving them; by decreasing or increasing the intensity of religion; by scientific and cultural exchanges and friendship tours, or by increased isolation; by appeasement, or early confrontation, or persuasion, by economic boycotts or trade agreements; by improving education or increasing the food supply; by fanning the flames of patriotism or by dousing the excesses of nationalism, or by any other possible means.

But if all fails and the armies of evil again stalk the earth, the forces of good must be willing, ready, and able to march forth and destroy the enemy with the utmost resolution, courage, and dispatch.

Bibliography

WORLD WARS I AND II – CANADA

Bell, K., and C.P. Stacey. *Not in Vain*. Toronto: University of Toronto Press, 1973.

Berton, P. *Vimy*. Toronto: McClelland and Stewart, 1986.

Broadfoot, B. *Six War Years 1939–1945*. Toronto: Doubleday, 1974.

Brock, J.V. *The Dark Broad Seas*. Toronto: McClelland and Stewart, 1981.

Churchill, W.S. *The World Crisis*. 6 vols. New York: Scribners, 1923–31.

Heaps, L. *The Grey Goose of Arnhem*. Markham, ON: Paperjacks, 1977.

Hogg, I.V. *Gas*. Toronto: Ballantine, 1975.

Lamb, J.B. *The Corvette Navy*. Toronto: Macmillan, 1979.

– *On the Triangle Run*. Toronto: Macmillan, 1986.

Lawrence, H. *A Bloody War*. Toronto: Macmillan, 1979.

Macpherson, K.R. *Canada's Fighting Ships*. Toronto, Samuel Stevens, 1975.

McElheran, Irene Brock. *That's What I'm Here For*. Toronto: Ryerson Press, 1955.

McWilliams, J.L., and R.J. Steel. *The Suicide Battalion*. Edmonton: Hurtig, 1978.

Milner, M. *North Atlantic Run*. Toronto: University of Toronto Press, 1985.

Schull, J. *The Far Distant Ships*. Ottawa: Queen's Printer, 1961.

Stafford, D. *Camp X*. Toronto: Lester & Orpen Dennys, 1986.

Swettenham, J. *Canada and the First World War*. Toronto: McGraw-Hill Ryerson, 1973.

– *Canada's Atlantic War*. Toronto: Samuel Stevens, 1979.

Wood, H.F., and J. Swettenham. *Silent Witnesses*. Toronto: Hakkert, 1974.

WORLD WAR I – SEA, AIR, AND LAND

Barclay, C.N. *Armistice, 1918*. New York: Barnes, 1969.

Bennett, G. *Naval Battles of the First World War*. London: Pan Books, 1968.

Churchill, W.S. *The World Crisis*. 6 vols. New York, 1923–31.

Coombs, R.E.B. *Before Endeavors Fade*. London: Battle of Britain Prints, 1976.

Cooper, B. *The Ironclads of Cambrai*. London: Pan Books, 1967.

Costello, J., and T. Hughes. *Jutland 1916*. London: Futura, 1976.

Dictionary of Disasters at Sea during the Age of Steam, 1824–1962. 2 vols. London: Lloyds Register of London, 1969.

Farrar-Hockley, A.H. *The Somme*. London: Pan Books, 1966.

Fredette, R.H. *The Sky on Fire. The First Battle of Britain 1917–18*. London: Holt Rinehart, 1966.

Hodges, G. *Memoirs of an Old Balloonatic*. London: Kimber, 1972.

Hogg, I.V. *The Guns 1914–18*. London: Pan Books, 1971.

Hoyt, E.P. *The Zeppelins*. New York: Lothrop, 1969.

Keegan, J. *Opening Moves 1914*. London: Pan Books, 1973.

MacDonald, L. *The Roses of No Man's Land*. London: Papermac, 1984.

– *Somme*. London: Papermac, 1983.

– *They Called It Passchendaele*. London: Macmillan, 1983.

Macksey, K. *Vimy Ridge*. London: Pan Books, 1973.

McKee, A. *Vimy Ridge*. London: Pan Books, 1966.

Messenger, C. *Trench Fighting 1914–18*. London: Pan Books, 1973.

Middlebrook, M. *The First Day on the Somme*. New York: Norton, 1972.

Mr Punch's History of the Great War. London: Cassell, 1920.

Pill, Barrie, ed. *History of the First World War*, 3 vols. London: Imperial War Museum.

Simpson, C. *Lusitania*. New York: Penguin Books, 1974.

Stock, J.W. *Zeebruge and Ostend*. New York: Random House, 1974.

Swettenham, J. *To Seize the Victory*. New York: McGraw-Hill, 1965.

Toland, J. *No Man's Land*. New York: Random House, 1980.

Tuchman, B.W. *The Guns of August*. New York: Bonanza, 1982.

– *The Zimmermann Telegram*. New York: Random House, 1966.

Wolff, L. *In Flanders Fields*. New York: Random House, 1958.

Wren, J. *The Great Battles of World War I*. London: Castlebooks, 1971.

WORLD WAR II – SEA, AIR, AND LAND

Allen, H.R. *Who Won the Battle of Britain?* London: Granada, 1976.

Ambrose, S.E. *Pegasus Bridge, June, 1944*. New York: Pocket Books, 1985.

Auty, P. *Tito*. New York: Ballantine, 1972.

Barker, R. *Torpedo Bomber!* New York: Ballantine, 1967.

– *The Thousand Plan*. London: Pan Books, 1967.

Bekker, C. *The Luftwaffe War Diaries*. New York: Ballantine, 1966.

Bishop, E. *Their Finest Hour*. New York: Ballantine, 1968.

Botting, D. *In the Ruins of the Reich*. London: Grafton Books, 1985.

Bradford, E. *The Mighty Hood*. London: Coronet, 1961.

Brookes, E. *The Gates of Hell*. London: Arrow, 1973.

Bryant, B. *Submarine Commander*. New York: Bantam, 1980.

Busch, F. *The Sinking of the Scharnhorst*. London: Futura, 1974.

Busch, H. *U-Boats at War*. New York: Ballantine, 1955.

Caidin, M. *The Night Hamburg Died*. New York: Ballantine, 1960.

Calder, A. *The People's War – Britain 1939–45*. London: Jonathan Cape, 1969.

Campbell, J. *The Bombing of Nuremberg*. London: Futura, 1974.

Carell, P. *Invasion – They're Coming!* New York: Bantam, 1964.

Charlwood, D. *No Moon Tonight*. London: Goodall, 1984.

Churchill, W.S. *The Second World War*. 6 vols. Boston: Houghton Mifflin, 1948–54.

Collier, R. *The Sands of Dunkirk*. New York: Dell, 1961.

Costello, John. *Love, Sex and War, 1939–1945*. London: Pan Books, 1986.

Cox, R. *Operation Sealion*. London: Arrow Books, 1975.

Creighton, K. *Convoy Commodore*. London: Kimber, 1956.

Deighton, Len. *Fighter – The True Story of The Battle of Britain*. London: Jonathan Cape, 1977.

Fitzgibbon, C. *The Blitz*. London: Corgi Books, 1974.

Forester, C.S. *Hunting the Bismark*. London: Mayflower Books, 1970.

Frank, Anne. *Diary of Anne Frank*. New York: Pocket Books, 1953.

Frank, W., and B. Rogge. *The German Raider Atlantis*. NY: Ballantine, 1956.

Gilbert, M. *Churchill: A Life*. New York: Holt, 1991.

Green, W. *The Observer's Book of Aircraft*. London: Warne, 1965.

Gretton, P. *Convoy Escort Commander*. London: Transworld Publishers, 1971.

Hamilton, N. *Monty*. Toronto: Fleet Books, 1982.

Hampshire, A.C. *The Secret Navies*. London: Sphere, 1980.

Hart, B.H.L. *The Other Side of the Hill*. London: Pan Books, 1983.

Hastings, M. *Bomber Command*. London: Pan Books, 1981.

Hersey, J. *Here to Stay*. New York: Bantam, 1963.

– *The Wall*. New York: Knopf, 1950.

Hillary, R. *The Last Enemy*. London: McMillan, 1942.

Hitler, A. *Hitler's Secret Conversations*. New York: Signet Books, 1961.

Howarth, D. *Dawn of D-Day*. London: Fontana/Collins, 1961.

Hoyt, E.P. *U-Boats Offshore*. New York: Stein & Day, 1978.

Jackson, R. *Dunkirk*. London: Mayflower Books, 1978.

Johnson, David. *The City Ablaze*. London: Kimber, 1980.

Jones, Ira. *Tiger Squadron*. New York: Award Books, 1954.

Kent County Council. *The County Administration in War 1939–45*. Maidstone: Kent County Council, 1946.

Lamb, G. *To War in a Stringbag*. New York: Bantam, 1980.

Langer, William, ed. *An Encyclopedia of World History*. Boston: Houghton Mifflin, 1940.

Leggett, E. *The Corfu Incident*. London: New English Library, 1976.

Lewin, R., ed. *The War on Land*. London: Arrow Books, 1972.

Lochner, L.P., ed. *The Goebels Diaries*. New York: Award Books, 1958.

Macintyre, D. *Aircraft Carrier*. New York: Ballantine, 1968.

Macksey, K. *Military Errors of World War II*. Toronto: Steddart, 1987.

Marshall, S.L.A. *Night Drop*. New York: Bantam, 1962.

Mason, David. *Raid on St Nazaire*. London: Macdonald, 1970.

– *Salerno*. New York: Ballantine, 1972.

Middlebrook, M. *Convoy*. NY: Morrow, 1977.

Millar, G. *The Bruneval Raid*. London: Pan Books, 1976.

Nekrasov, V. *Front-line Stalingrad*. Canada: Fontana/Collins, 1964.

Neumann, P. The Black March. New York: Bantam, 1960.

Nytrup, P. *An Outline of the German Occupation of Denmark 1940–1945*. Museum of the Danish Resistance Movement, 1968.

Osment, H. *Wartime Sherborne*. Sherborne: Dorset Publishing Co., 1984.

Pack, S.W.C. *The Battle of Matapan*. London: Pan Books, 1961.

Peillard, L. *The Laconia Affair*. Toronto: Bantam, 1961.

– *Sink the Tirpitz*. London: Granada, 1975.

Poolman, K. *Ark Royal*. London: New English Library, 1975.

Porten, E.P. von der. *The German Navy in World War Two*. New York: Ballantine, 1974.

Prien, N.G. *U-Boat Commander*. New York: Award, 1976.

Rayner, D.A. *Escort*. London: Futura, 1974.

Reid, P.R. *Colditz*. London: Coronet Books, 1962.

Richards, D., and H. Saunders. *Royal Air Force 1939–1945*. 3 vols. London: Her Majesty's Stationery Office, 1953–54.

Roskill, S.W. HMS *Warspite*. London: Futura, 1974.

– *The Navy at War*. London: Collins, 1960.

Ryan, C. *The Last Battle*. New York: Pocket Books, 1967.

– *The Longest Day. June 6, 1944*. New York: Crest Books, 1959.

Shankland, P., and A. Hunter. *Malta Convoy*. London: Fontana, 1963.

Shirer, W.L. *End of a Berlin Diary*. New York: Popular Library, 1961.

Sims, E.H. *The Greatest Aces*. New York: Ballantine, 1970.

Speidel, H. *Invasion 1944*. New York: Paperback Library, 1968.

Steiner, J. *Treblinka*. New York: Simon & Schuster, 1967.

Swenson, A. *The Raiders: Desert Strike Force*. New York: Ballantine, 1968.

Thomas, Ben. *Ben's Limehouse*. London: Ragged School Books, 1987.

Thomas, G., and M. Witts. *Guernica*. New York: Ballantine, 1975.

Thompson, R.W. *Spearhead of Invasion D-Day*. London: Ballantine, 1968.

"The Town with an Educated Heart." *Readers' Digest* (November 1943).

Tubbs, D.B. *Lancaster Bomber*. London: Ballantine, 1972.

Tute, W., J. Castello, and T. Hughes. *D-Day*. London: Pan Books, 1975.

Warren, C.E.T., and J. Benson. *The Midget Raiders*. New York: Morrow, 1968.

Werner, H.A. *Iron Coffins*. New York: Kinney, 1971.

Whiting, C. *Bounce the Rhine*. London: Grafton Books, 1985.

Williams, J. *France, Summer 1940*. London: Macdonald & Co., 1969.

Woodward, D. *The Tirpitz*. New York: Berkley, 1963.

Young, E. *One of Our Submarines*. London: Pan Books, 1968.

Young, Peter. *Commando*. London: Ballantine, 1969.

INTELLIGENCE

Beesly, P. *Very Special Intelligence*. New York: Ballantine, 1977.

Bittman, L. *The Deception Game*. New York: Ballantine, 1972.

Brook-Shepherd. *The Storm Petrels*. New York: Ballantine, 1977.

Calvocoressi, P. *Top Secret Ultra*. London: Sphere Books, 1980.

Castle, J. *The Password is Courage*. New York: Ballantine, 1954.

Collins, L., and D. Pierre. *Is Paris Burning?* New York: Pocket Books, 1965.

Farago, L. *The Game of the Foxes*. Toronto: Bantam, 1971.

Fisher, D. *The War Magician*. New York: Berkley Books, 1983.

Fourcade, M. *Noah's Ark*. New York: Ballantine, 1975.

Hagen, L. *The Mark of the Swastika*. New York: Bantam, 1965.

Harris, L., and B. Taylor. *Escape to Honor*. Toronto: Paperjacks, 1985.

Hawes, S., and R. White, eds. *Resistance in Europe: 1939–45*. Harmondsworth: Penguin, 1976.

Jones, R.V. *Most Secret War*. London: Coronet Books, 1979.

Kahn, D. *Hitler's Spies*. New York: Macmillan, 1978.

Leasor, J. *Code Name Nimrod*. New York: Playboy Paperbacks, 1980.

Lewin, R. *Ultra Goes to War*. New York: McGraw-Hill, 1978.

McDougall, M.C. *Swiftly They Struck*. London: Grafton/Collins, 1954.

Masterman, J.C. *The Double Cross System*. Yale University Press, 1972.

Montagu, E. *Beyond Top Secret U*. London: Corgi Books, 1979.

– *The Man Who Never Was*. New York: Lippincott, 1966.

Moravec, F. *Master of Spies*. London: Sphere Books, 1981.

Mure, D. *Phantom Army*. London: Sphere Books, 1979.

Pawle, G. *Secret Weapons of World War II*. New York: Ballantine, 1968.

Prijol, J. *Garbo*. London: Grafton Books, 1986.

Rachlis, E. *They Came to Kill*. New York: Poplar, 1962.

Read, A., and D. Fisher. *Operation Lucy*. New York: Coward, 1981.

Reit, S. *Masquerade*. New York: Hawthorne Books, 1978.

Richardson, A. *Wingless Victory*. London: Pan Books, 1956.

Stevenson, W. *A Man Called Intrepid*. New York: Harcourt, 1976.

– *Intrepid's Last Case*. New York: Ballantine, 1984.

West, N. *MI5*. London: Triad, 1983.

Wright, P. *Spy Catcher*. New York: Viking, 1987.

V-1S AND V-2S

British Information Services. *Flying Bombs, December 1944*. New York: Agency of the British Government. Pamphlet.

Cookley, P.G. *Flying Bombs*. New York: Scribner, 1979.

Illingworth, F. *Flying Bomb: Story of the V-1 and V-2*. Victoria, Egham: Citizen Press, 1945.

National Association of Spotter's Club. *Flying Bomb*. Vol. 15. London: Rolls House, English Catalogue Books, 1942–47.

Young, Richard A. *The Flying Bomb*. London: Ian Allen, 1978.

Index